# THE WORLD OF NUMBERS

# THE
# WORLD OF NUMBERS

BY

## HERBERT M<sup>C</sup>KAY

CAMBRIDGE

*At the University Press*

1946

# CAMBRIDGE
## UNIVERSITY PRESS

University Printing House, Cambridge CB2 8BS, United Kingdom

Cambridge University Press is part of the University of Cambridge.

It furthers the University's mission by disseminating knowledge in the pursuit of education, learning and research at the highest international levels of excellence.

www.cambridge.org
Information on this title: www.cambridge.org/9781107494992

© Cambridge University Press 1946

First published 1946
First paperback edition 2015

*A catalogue record for this publication is available from the British Library*

ISBN 978-1-107-49499-2 Paperback

# CONTENTS

*Preface*                                                    page vij

*Chapter* 1.  A Girdle round about the Earth                    1

2.  The Trials of Geographers                    18

3.  The Arithmetic of Great Rivers               29

4.  Widdershins                                  42

5.  The All-composing Hour                       48

6.  Ellipses and Orbits                          63

7.  Proportions in the Stars                     78

8.  The Earth in a Balance                       94

9.  The Prodigal Sun                            114

10.  Magnitudes of the Stars                    124

11.  Symbols                                    131

12.  Squaring the Circle                        143

13.  The Ghost Quantity                         150

14.  Poets' Numbers                             161

15.  Numbers that mean too much                 176

16.  Was Man made by the Earth?                 187

# PREFACE

## *The Mathematical Picture*

Mathematics is a way of looking at things. No one pretends that it is the only way, but it is an important way. Those who have failed to cultivate the mathematical outlook are missing a good deal, even though they may be unaware of what they are missing. Just as a colour-blind man is unaware of the colours denied to him. The ignorant or unfortunate live in a less interesting world than those who use all their talents. Nearly all intelligent people take an interest in the universe. We want to know about the stars: their distances and movements, their constitution and conditions. We would like to know far more than we do about our nearest neighbours, the moon and the planets. Well, our knowledge of things outside the earth, as well as of a great many things on the earth itself, is almost entirely mathematical, or at least based on mathematics.

We can satisfy a rational curiosity about these things, either by accepting uncritically the statements of astronomers and others, or by working things out for ourselves and fitting them into a reasoned mathematical picture of the conditions in which we live. So long as we do not aim at technical accuracy, the mathematics is quite simple, no more than a great many people take with them when they leave school.

I have to thank several people for assistance in checking the proofs, and especially Stevens of the Upper Sixth who nobly checked all the mathematics.

H. M.

1944

# CHAPTER 1

## *A Girdle round about the Earth*

PEOPLE SOMETIMES imagine mathematicians sitting in armchairs, with heavy tomes about them, solemnly working out abstruse problems of no interest to anyone but themselves. Calm, dispassionate, there they sit, thinking only of numbers and the world of numbers. Dispassionate, for who could be passionate about mere numbers?

Not a bit of it.

People who try to find humdrum explanations of things that are not humdrum, have tried to explain the joy of mathematics by saying that it consists of the mental satisfaction of getting sums right. It makes me laugh, as Herodotus exclaimed two thousand years ago. He knows little of mathematics who thinks that; because mathematics is of all subjects the most adventurous. It adventures alike into the infinitely small and the infinitely great; it is the one subject that can stretch out confidently to the bounds of the universe.

And at any moment new and unsuspected mathematical horizons may open up. Even to follow in the footsteps of the great discoverers is exciting enough. To follow de Moivre through the steps of his exciting theorem, to be suddenly confronted with the whole new world of imaginary trigonometry. To sit with Newton and watch the fall of the famous apple that fired his imagination with the idea that falling apple and falling moon obey a single universe-wide law. With Cavendish to set out to weigh the world in a balance.

In the year 1500 a coach rolled out of Paris on a north and south road. M. Fernel leaned out of the coach with his eyes fixed on one of the carriage wheels. People stared at him; small boys shouted ribald remarks; but his eyes remained serenely glued to the wheel. And his lips moved as he counted, and went on counting. What was this extraordinary man doing? Strange as it may sound, he was measuring the earth. And lest anyone else should think he was as ridiculous as the ribald youths seemed to think, I hasten to add that the result of his measurement was quite good.

After all, since men began to think, who had not wondered about the size of the earth, even if only to dismiss the idea of measuring it as an impossibility? 'Who hath stretched a line upon the earth?' was demanded of Job. There was no answer. It must have seemed an impossible task, so long as the idea of a fixed earth persisted.

It is the rotation of the earth that makes its measurement a comparatively simple operation, and that justified M. Fernel in counting the turns of his carriage wheel. Rotation provides us with a natural framework on which we can make the necessary measurements. It provides us with two fixed points; these are the North Pole and the South Pole, the ends of the axis on which the earth rotates. It provides us also with a fixed circle; this is the equator, the great circle halfway between the poles.

We use the equator as a zero line, and we measure latitudes north and south from it. Measurements of latitude are made in degrees, for the simple and satisfactory reason that it is much easier to measure angles than to measure lengths. The mere turn of a telescope on a graduated circle is usually sufficient to measure an angle; whereas the measurement of a considerable length is a difficult process. Very often indeed a direct measurement of length is impossible. Navigating officers at sea may guess their positions by dead reckoning, based on vague measurements of distance; but after a period of clouded skies they must long for a sight of the punctual stars with their promise of exact angular measurement.

In measuring a length on land we have to make allowances for slopes and inequalities, so that it is advisable to choose the flattest piece of land available. No one, I should imagine, would choose the Himalayas for such a measurement, though on one famous occasion the incurably romantic spirit of French mathematicians drove them to mountainous Peru. Commonsense Englishmen chose a level tract on Salisbury Plain to make an exact measurement of length; Russian mathematicians are fortunate in having vast level plains for their measurements; a north and south line through Paris is level enough; and there are few countries without level stretches to simplify their measurements of length. Indeed, the modern method of making all measurements from a single carefully measured base originated on the frozen meadows of Holland.

In measuring a straight line we have to make sure that the measuring rods are evenly spaced and in an exactly straight line; we have to make allowances for the imperfections of the measuring rods, and especially for their expansion by heating; we have to make allowance for the curvature of the earth on which we are measuring, so that the line may deviate from a strictly straight direction to follow the curvature. The difficulty of measuring a long line following the slight curve of the earth, with anything approaching exactitude, is so great that it is usual to measure one line a mile or more in length, and then to make all the other measurements in angles. The original base line, measured with extreme care, enables all the other lengths to be calculated, by the methods of trigonometry, from the measured angles.

Before we begin to attempt to measure the earth we have to have some idea of its shape. We are measuring, as it were, blind-fold; we can see nothing of the actual shape of the world, and very little even of the lines we are measuring. I find it difficult to imagine how a Flat Earthist would set about the job of measuring the earth. Random measurement would take us no-where in particular; so that even a Flat Earthist must make some kind of assumption about his flat earth before he can begin to measure it.

It is not unreasonable to suppose that the earth is spherical in shape; there have always been the round sun and the equally round moon to suggest such a shape. If we decide that the earth is spherical, we have got something to go on. We might, for example, find two places, A and B, one of which is exactly north of the other. We measure the length AB by the best means at our disposal; we may, with great labour, measure the distance with measuring rods, or we may pace it, or we may rely on the 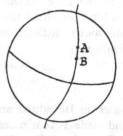 estimates of travellers. Then we need to find the latitudes of A and B, that is, their angular distances from the equator. Suppose the difference in latitude is 4°; the distance *AB*, which we have measured, represents 4°. All we have to do is to find what distance represents 360°; that is, we multiply the distance *AB* by $\frac{360}{4} = 90$.

The ancient Greeks had clear ideas about the shape of the earth; they imagined it to be spherical, and so big that the heights of mountains made no appreciable difference to the rotundity. Further, they were able to measure latitudes by observations of the sun and the stars. Eratosthenes, who flourished, as one says, about 200 B.C., made a measurement of the circumference of the earth; he gave it as 252,000 stadia. A stadium was the length of the track for foot-races; we are more familiar with the word as the name for the terraced seats for spectators round the track— this was a later use of the word. Measurement of the stadium at Athens has shown that the length was 607 feet, so that Eratosthenes' calculation was:

$$252,000 \times 607 \text{ feet} = \frac{252,000 \times 607}{5280} \text{ miles}$$
$$= \text{about } 29,000 \text{ miles.}$$

When we consider the difficulty of measuring latitudes and long distances without the use of precision instruments, that is an extremely good approximation to the circumference of the earth. Eratosthenes got his difference of latitude in a rather extraordinary way. At Syene in Upper Egypt there was a deep well. On one day of the year, the summer solstice, when the sun reaches its highest point in the sky, there was no shadow at the bottom of the well at noon. The sun must then have been exactly overhead. At noon on the same day the sun at Alexandria was 7° 12 from the zenith; so the difference in latitude was 7° 12′. The north and south distance was 5000 stadia, so Eratosthenes had the simple sum:

$$5000 \times \frac{360°}{7° \ 12'} = 5000 \times 50$$
$$= 250,000 \text{ stadia.}$$

Later on he added another 2000 stadia to his calculated result and made it that much less accurate.

Eratosthenes' method was correct, and so also was his arithmetic. But he had to rely on the estimates of surveyors for the distance between Syene and Alexandria, and they made the distance about 25 per cent greater than we now know it to be.

Posidonius came later with an estimate of 240,000 stadia, which accidentally came nearer the truth. He too revised his estimate. Unfortunately he reduced it to 180,000 stadia, which is just over 20,000 miles. Unfortunately also the great Ptolemy accepted this shrunken estimate; and the world remained shrunken in men's minds, at any rate in the minds of those who thought about such things, until Columbus startled them into a realisation of the neglected 5000 miles of circumference.

After the Greeks—eclipse. It was not till long afterwards that the Arabs took up the tale. About the year A.D. 800 the Caliph Almamoun decided that the earth should be measured. He carried out his project in a spectacular way worthy of a great caliph. He assembled two parties of astronomers back to back on the plains of Mesopotamia where there were no mountains to interfere with the measurements. The two parties measured the latitude of the starting-point; and then they set off in opposite directions, measuring the distance with rods as they went. One party went north and the other south, and they continued until each had reached a point one degree from the zero point. The double measurement gave them the length of two degrees of latitude, and they had then only to multiply this length by 180 to find the circumference of the world. But it was not an easy task. It is far from easy to measure a distance of about seventy miles with rods, and to keep the measurements in anything like a straight line; it is far from easy to measure exact latitudes with rude instruments. The wily caliph ensured that there should be no 'distribution of errors' or other cooking of results, by segregating the two parties. There was a distance of 140 miles between them when they had done measuring, and the lack of collaboration produced poor results. He was probably a saddened and disillusioned caliph with a much shaken faith in astronomy and mathematics.

Another long eclipse. And then in 1500 Fernel rolled out of Paris in his coach, counting the number of turns of one of the wheels as he went. He had only to measure the circumference of the wheel and multiply it by the number of turns to get a very fair estimate of the distance. As the road ran north and south, he had measured part of a meridian. By a bit of luck he got a good angular measurement too; so he did get a fair result.

In 1581 an infant prodigy was born at Leyden in Holland; he was Willebrod Snell. By the age of twelve he seems to have mastered the mathematics of his day; and he sought other fields. There were plenty around him. The frozen meadows of Holland gave him level ground for a good base line; and he initiated there the modern method of triangulation. He measured a base line, and then found the angular distances of points around it. The rest is a matter of trigonometrical calculation. The calculations are not easy, because allowances have to be made for the curvature of the earth, for ups and downs, and for small inaccuracies of measurement. But they can be made with great accuracy. Snell's innovation made it possible to measure considerable arcs of meridians, and of parallels of latitude, with greater accuracy than was possible, and less labour than was necessary, before his time.

That is just about as far as it was possible to get in earth measurement; or so it seemed. There might be a little increase in exactness, and that is all: from a vague measurement to the nearest thousand miles, then to the nearest hundred, and so on with progressive increases in exactness. So it seemed. Until a French astronomer named Richer went to the island of Cayenne in Guiana. He took with him a clock which accurately beat seconds in Paris. But in Cayenne it went 2½ minutes slow every day. Most travellers would have shortened the pendulum and thought no more about it. Not so an astronomer. Especially when other astronomers, adventuring toward the equator, found the same thing.

There had to be a reason, and it was Newton who saw the reason. The clock went slow because the pendulum fell more slowly than when it was in Paris. That is, the acceleration due to gravity was less, and this could only mean that gravity itself was less. The pendulum actually weighed less near the equator than when it was farther north. That could be due to one of two things, or to both in combination. We might be farther from the centre of the earth at the equator, in which case gravity would be less; or the greater rate of rotation at the equator would increase outward 'centrifugal force' and so ease the downward pressure that is weight. Newton deduced, from the facts, that the earth had an equatorial bulge, and a consequent flattening at the poles.

The earth had ceased to be a sphere. That is of course the conception of the earth in the minds of mathematicians, the conception on which measurements of the earth are made. A fresh model of the earth had to be found.

If we cut out such ridiculous shapes as approximate cubes and pyramids, and stick to the models we have almost constantly before our eyes—the sun and the moon, and when we look through a telescope, the planets; when we are thus reasonable we have a choice between three shapes.

One of these shapes is the lemon shape, spinning round the long axis down the middle; this shape is called a prolate spheroid (I). If we cut across it at the equator, or anywhere parallel to the equator, the section is a circle. But if we cut through a meridian (*ABC*) the section is an ellipse with the axis of rotation (*AC*) as the major axis. Also, if we mark the spheroid with degrees of latitude from the equator to the poles, the spaces between the parallels decrease toward the poles; that is, the length of a degree of latitude decreases toward the poles.

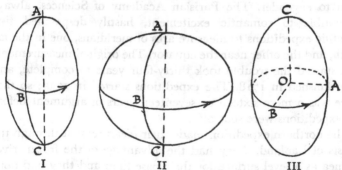

I          II          III

Another possible shape is the orange shape, spinning round on its short axis; this shape is called an oblate spheroid (II). Again, if we cut through parallel to the equator, the sections are circles; but if we cut through a meridian (*ABC*) the section is an ellipse with the axis of rotation (*AC*) as its minor axis. Degrees of latitude are longer toward the poles than near the equator.

The third possible shape is an ellipsoid (III). This shape differs

from the spheroids in having sections parallel to the equator elliptical, and not circular; so that there are two unequal axes, or radii
(*OA* and *OB*), at the equator. A section through a meridian
would also be an ellipse, with *OC* as one of the axes.

The first choice was between the spheroids. After all, the earth
is rotating, and we can form a spheroid by rotating an ellipse
round one of its axes; and it is highly probable that rotation will
spread matter evenly round the axis, that is, in circles, and not in
ellipses. Newton and Huyghens favoured the oblate spheroid,
partly because of the behaviour of pendulums that were taken to
the tropics, and partly because rotation would swing matter
outward at the equator in the days when the earth was plastic.

There seems to have been general agreement about the oblate
spheroid; it was a reasonable idea. And then, about 1700, the
Cassinis, father and son, measured an arc of the meridian through
Paris. The measurement showed that a degree in the northern
part was shorter than a degree in the southern part. It seemed
that the earth was a prolate spheroid. Sensation. The scientific
world seethed with excitement. Prolate or oblate? The question
was argued with all the bitterness that mathematical disputes
seem to engender. The Parisian Academy of Sciences, always
susceptible to romantic excitement, hastily despatched two
scientific expeditions to measure arcs of meridians, one in the far
north, and the other near the equator. The original measurements
that started the trouble took thirty-four years to complete, and
were finished in 1718. The expeditions started in 1735; so that
there was a mere sixteen or seventeen years of argument before
the expeditions were sent off.

The northern expedition made their measurement through the
forests of Lapland. They had the advantage of the frozen river
Tornea as a level surface for their base line; and they had done
with the line before a thaw came. They carried their triangulation
north and south from the base line, and succeeded in measuring
very nearly a degree of latitude. Except that they were frost-
bitten in winter, and eaten up by flies and mosquitoes in summer,
this expedition had a comparatively easy time. They returned
with their measurement, the length of a degree of latitude near
the Polar Circle, after sixteen months.

The equatorial expedition went to Quito in Peru, and there they stayed for ten years. They measured an arc of three degrees, over mountain and valley, with one observation post down in the valley, and the next post perhaps half a mile higher up on a mountain slope. Some of the instruments turned out to be unreliable and had to be readjusted and improved. The natives stared at the scientists in amazement and then in alarm; they had no doubt that something malign was intended toward them, perhaps the filching of their lands; telescopes can look uncomfortably like some new kind of firearm. The natives became openly hostile, and attempted to slaughter the scientists. The scientists fought them off, and continued to measure. As if that were not enough, internecine conflict broke out between the two leaders of the expedition. They would not even measure together. Each went haughtily his own way, made his own measurements, and wrote his own book about the expedition.

It is worthy of mention that the measurements were made in toises. Now there were various toises of different lengths, but after that great mathematical adventure—for all the world like one of Jules Verne's stories—a toise was emphatically 'the toise of Peru', which is 6 pieds, or 2·1315 yards.

After that, who shall say that mathematicians are not romantic?

I almost forgot to say that the result of the two expeditions was to show that a degree of latitude near the Arctic Circle is definitely longer than a degree at the equator. The oblate spheroid had it.

And after all the excitement the line measured by the Cassinis was remeasured with more care, and it turned out that the Cassinis were wrong. Why was not that done at first? you may ask. As I have shown, the answer lies in the excitable romantic natures of mathematicians, and especially of French mathematicians. They carried out the job in the grand manner.

More commonplace, but never quite commonplace, surveys have since been made. Even in the small compass of the British Isles there was room for a measurement that embraced Snowdon in Wales, Slieve Donard in Ireland, and Sca Fell in Cumberland as the apices of a single triangle, each side of which is over a hundred miles long.

Surveys in all parts of the world amply confirm that the earth is an oblate spheroid. Measurements of arcs of meridians fit the equation:

$$\frac{x^2}{3963\cdot3^2}+\frac{y^2}{3950^2}=1,$$

which is the equation of an ellipse. That is, the equation which best satisfies the various measurements. It shows that a section through a meridian is an ellipse whose major and minor axes are 3963·3 miles and 3950 miles; the denominators are found in miles. The idea of an ellipsoid has not been neglected; but if the equatorial section is an ellipse, it is so nearly circular as to make little difference.

Two measurements are all that we need to describe an oblate spheroid completely. If we know the lengths of the semi-axes $OA$ and $OB$ we can construct the ellipse, which is a section through a meridian. We have only to rotate the ellipse about the minor axis $BB_1$ to complete the spheroid. The lengths of the axes cannot of course be measured directly. They have to be calculated from the 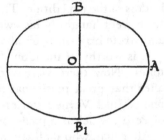 lengths of the measured meridian arcs, and the lengths of parallels of latitude.

However, the shape of the earth is summed up in the two measurements:

polar radius:      3950 miles,<br>
equatorial radius: 3963⅓ miles.

The difference caused by the polar flattening is 13⅓ miles; an observer at sea-level at one of the poles would be 13⅓ miles nearer the centre of the earth than a similar observer at the equator. We are interested not so much in the actual difference in miles, as in what fraction this is of the equatorial radius. It is this fraction that determines how much the shape of the earth differs from a sphere. The fraction is called the *ellipticity*.

Ellipticity of the earth $=\dfrac{13\frac{1}{3}}{3963\frac{1}{3}}=\dfrac{1}{297}$, and that is a very little more than $\frac{1}{300}$, which is a third of one per cent.

The earth is circular at the equator, or at least so nearly circular that the difference is insignificant. The length of the equatorial circumference is therefore:

$$2 \times 3.1416 \times 3963\tfrac{1}{3} = 24,902 \text{ miles.}$$

This is commonly given as 25,000 miles, or more accurately as 24,900 miles.

Airmen have claimed to have flown round the world, and that is a very dubious statement. To fly along, or close to, a parallel of latitude would only be a true circumnavigation at the equator. At 50° latitude the length of a degree of longitude is only 44·54 miles, so that the length of the whole parallel of latitude is only 44·54 × 360 = just over 16,000 miles or less than two-thirds of the equator. And of course the distance fades away to nothing at the poles. At the poles the distance round the world, interpreted in the sense of the length of a parallel of latitude, is zero; and that reduces that sense to nonsense.

Magellan and Drake were true circumnavigators. Their voyages round the world were not far from the equator, with detours where these were rendered necessary by the configuration of the continents.

We have ignored the irregular differences made by the heights of mountains. The Greeks saw long ago that these heights were insignificant compared with the radius of the earth. The highest mountain is considerably higher than any they knew, but even the $5\tfrac{1}{2}$ miles of the Himalayas is a mere $\dfrac{5\tfrac{1}{2}}{3963\tfrac{1}{3}} = \dfrac{1}{720}$ of the equatorial radius.

Even the polar flattening is not much more than $\tfrac{1}{300}$, so that the earth can be represented as an exact sphere, with no error that could be detected by eye. In a globe of 2 feet radius the polar flattening would be:

$$\tfrac{24}{300} \text{ inch} = \text{about } \tfrac{1}{12} \text{ inch.}$$

The height of the highest mountain would be:

$$\tfrac{24}{720} \text{ inch} = \tfrac{1}{30} \text{ inch.}$$

There is nothing to make it worth while departing from the spherical shape.

We are apt to be a bit superior about our knowledge of the shape of the earth, even though we still have our Flat Earthists. But really there was every excuse for landsmen believing that the world is flat. All the landsman sees is the unevenness of hill and dale that disguises the rotundity. Only the seaman with his constant view of the circular horizon, and the astronomer watching the curved shadow of the earth on the moon, must always have had inklings of the spheroidal truth. But even such evidence, it was not beyond the ability of sophistry to disparage and explain with some degree of plausibility.

There is a well-known geometrical theorem that enables us to find the curvature of the earth very easily and with fair accuracy. In the diagram $AB$ is a diameter of the earth, and $CD$ is a chord crossing it at right angles, so that the parts marked $a$ are equal. The theorem is:

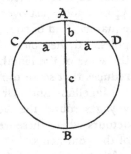

$$a^2 = bc.$$

Suppose we want to find the curvature across the distance of a mile, we make $a = \frac{1}{2}$ mile, and we want to find $b$, the hump as it were. $b + c$ is the diameter of the earth, and $b$ is so small compared with $c$ that we say $c = 8000$ miles.

Then:
$$a^2 = (\tfrac{1}{2})^2 = \tfrac{1}{4},$$
$$bc = 8000b,$$
$$8000b = \tfrac{1}{4},$$
$$b = \frac{1}{4 \times 8000} \text{ mile}$$
$$= \frac{5280 \times 12}{4 \times 8000} \text{ inches}$$
$$= 1 \cdot 98 \text{ or } 2 \text{ inches.}$$

The curvature of the earth's surface, the upward bulge, is a bare 2 inches in a mile. It is little wonder that the idea of a flat earth lingered on for many centuries after the first enlightened philosophers had imagined the earth to be a sphere.

The diagram we have just been looking at illustrates an interesting point about tunnels. We may take the chord to represent a tunnel. This tunnel is 20 miles long, and the engineers have taken extreme care to cut it in an exactly straight line. That was unfortunate. To find the length $b$ we have:

$$bc = a^2,$$
$$8000b = 10^2 = 100,$$
$$b = \tfrac{100}{8000} \text{ mile}$$
$$= \tfrac{528}{8} \text{ feet}$$
$$= 66 \text{ feet.}$$

The centre of the tunnel is therefore 66 feet nearer the centre of the earth than the ends; that is to say, it is 66 feet lower than the ends. Water would run downhill from both ends, and accumulate at the centre. We also have the apparently odd fact that a train, running in a straight line without changing direction, would run downhill during the first half of the journey and uphill during the second half. The reason is that 'down' is not a direction like east or west, though we usually treat it as such; but rather like north or south. That is to say it is toward a definite point, the centre of the earth. When we pass a point, either through it or at a distance, we always approach it and then recede, without the necessity of changing direction. Just as a polar explorer might go north to the North Pole and continue south without changing direction. But east is east, and west is west.

If we want the tunnel to be free from water we have to see that the middle rises rather more than 66 feet above the ends. That would provide a fairly level track for the trains; an exactly level track would follow the curvature of the earth.

If we are creeping along the earth's surface mile by mile we have nothing to worry about in the way of curvature; the recurring 2 inches is neither here nor there. Indeed, when we want to say a thing is very simple, we say it is 'plane sailing'. The expression comes from the practice of mariners who sail on narrow seas. They make their calculations as if the sea were flat—they are the true Flat Earthists. And as we have seen, they are not likely to go far wrong. Plane sailing is adequate for narrow seas.

But sailors who navigate the great seas have to allow for the curvature of the earth. Their voyages are not even approximately

straight lines, but distinct arcs of circles. Even in so short a
voyage as 400 miles, the length of England, we have:

$$a = 200 \text{ miles,}$$
$$8000b = 200^2,$$
$$b = \frac{200^2}{8000}$$
$$= 5 \text{ miles.}$$

Thus England has an upward bulge of 5 miles. The mid-point of
a straight line joining the north and south of England is 5 miles
below the ground. To treat England as a plane surface is like
ignoring a vast mountain 5 miles high.

When distances are greater than 400 miles we can no longer
ignore the fact that $c$ is not strictly equal to the diameter of the
earth. Across 2000 miles of the Atlantic Ocean or the American
Continent:

$$bc = a^2 = 1000^2$$
and
$$b + c = 8000 \text{ (the earth's diameter in miles);}$$
$$c = 8000 - b.$$
So
$$bc = b(8000 - b) = 1000^2$$
or
$$8000b - b^2 = 1000^2.$$
$$\therefore \ b^2 - 8000b + 1000^2 = 0.$$

An easy way of solving this equation is to write $b = 1000x$, and
then cancel $1000^2$.

$$1000^2 x^2 - 8000 \times 1000x + 1000^2 = 0,$$
$$x^2 - 8x + 1 \qquad\qquad = 0,$$
$$x \qquad = \frac{8 \pm \sqrt{64 - 4}}{2}$$
$$= 4 \pm \sqrt{15}.$$

The two values correspond to $c$ and $b$; there is nothing to
distinguish them in the equation. We want the smaller value:

$$4 - \sqrt{15} = 4 - 3 \cdot 873 = \cdot 127,$$
$$b = \cdot 127 \times 1000 = 127 \text{ miles.}$$

It is clear that the curvature of the earth's surface makes a
considerable difference in calculations of directions and distances.
Oddly enough we can find the general direction of a voyage
between two ports in the simplest possible way. On a globe
representing the earth we stretch a thread between the ports, and

the thread gives the shortest distance between the two ports. A thread can save a lot of calculation.

A thread stretched on the globe draws instant attention to the fact that the shortest line between two ports on an east and west line is not the parallel of latitude between them. It is part of the great circle between the two ports; if the thread is continued in the same direction it completes a great circle—the greatest possible circle that can be drawn on the globe.

There is another earth measure that has intrigued men from of old, the tilt of the earth's axis. We have all observed it; we cannot help observing it, even if only in the form of noting that the sun rises higher in the sky in June than in December. That is indeed what one observes.

Early astronomers observed that the sun reaches its highest point at the summer solstice (about 21 June), and that it reaches its lowest noon position at the winter solstice (about 22 December). The observations could hardly be missed by anyone making a systematic and continuous survey of the heavens. As long ago as 1100 B.C. careful measurements were made in China of the greatest and least heights of the noon sun. The difference was found to be $47° 48' 6·3''$. This difference we now ascribe to a tilt of half that angle in the earth's axis; that is, $23° 54' 3·15''$.

The observation, we say, could hardly be missed. Its interpretation was another and more difficult matter. It could equally well be interpreted in either of two ways.

There was the obvious interpretation that the sun in its yearly circuit of the heavens rises high toward the north at the summer solstice, and sinks deep toward the south at the winter solstice. The other interpretation was far from obvious. It is that the axis on which the earth rotates is tilted. When the sun is at the side of the earth toward which the axis tilts it appears high in the sky; when the tilt is away from the sun then the sun appears low in the sky. The actual tilt, according to this theory, is half the difference between the greatest and least heights of the sun.

The idea of a tilt of the earth's axis is now accepted as the better theory. The tilt toward the sun or away from the sun accounts for the changing seasons, and explains why northern summer coincides in time with southern winter.

In *Paradise Lost*, Milton, with fine impartiality, gives both theories. He describes the tilt of the axis, with all the vagaries it introduces into the climates of different regions, as part of God's punishment for man's first disobedience. In the passage quoted below 'twice ten degrees and more' is Milton's way of fitting in the tilt of $23\frac{1}{2}°$. The apparent path of the sun is described, north to the Tropic of Cancer in summer, and south to the Tropic of Capricorn in winter. The seven Atlantic sisters are the Pleiades, the daughters of Atlas; this star cluster is in Taurus, the Bull.

> Some say he bid his Angels turne ascanse
> The Poles of Earth twice ten degrees and more
> From the Suns Axle; they with labour push'd
> Oblique the Centric Globe: Som say the Sun
> Was bid turn Reines from th' Equinoctial Rode
> Like distant breadth to *Taurus*, with the Seav'n
> *Atlantick* Sisters, and the *Spartan* Twins
> Up to the *Tropic* Crab; thence down amaine
> By *Leo* and the *Virgin* and the *Scales*,
> As deep as *Capricorne*, to bring in change
> Of Seasons to each Clime; else had the Spring
> Perpetual smil'd on Earth with vernant Flours,
> Equal in Days and Nights, except to those
> Beyond the Polar Circles; to them Day
> Had unbenighted shon, while the low Sun
> To recompence his distance, in thir sight
> Had rounded still th' *Horizon*, and not known
> Or East or West.

One can differ from Milton as to the benefit of equal days and nights throughout the year. The rising and setting of the sun at six o'clock or so throughout the year is a monotonous feature of life in the Tropics, whereas the changing seasons bring variety into northern climates. The latitude of England is especially favoured with long summer days, without the penalty of perpetual night in winter.

That is a matter of opinion. Milton's error about the Polar Circles is not. The Polar Circles are $23\frac{1}{2}°$ from the Poles, and a direct result of the tilt of the axis; they are the parallels of latitude within which there is a 24-hour day in summer and a 24-hour night in winter. If there were no tilt of the axis, there would be no Polar Circles. People $23\frac{1}{2}°$ from one of the poles would have equal days and nights, with the sun rising to $23\frac{1}{2}°$ above the horizon, just as they do now at the spring and autumn equinoxes.

The conditions Milton describes apply to the actual poles only, and to regions round them. All directions from the North Pole are of course south, but the sun, rounding still the horizon, would be seen to move from east to west. He would be a poor Polarite who was not aware of east and west.

Since the poles *are* turned askance, an observer at the North Pole would see the sun appear on the horizon at the vernal equinox. He would see it spiral up the sky to a height of $23\frac{1}{2}°$ at the summer solstice, and then spiral down again, to reach the horizon at the autumnal equinox. The sun spirals up and down our own skies in the same way, except that the spiral is tilted, so that part of it is hidden beneath the northern horizon.

# CHAPTER 2

## *The Trials of Geographers*

THERE ARE a lot of things we want to know about the earth. We seem to have got reasonably accurate measurements of the equatorial and polar radii; so we should be able to answer questions about the area and volume of the earth.

The area is a comparatively simple matter, so long as we are satisfied with a fair degree of approximation, and do not try to push the matter too far. We shall not go far wrong if we assume the earth to be a sphere of radius 3960 miles. The area of such a sphere is:

$$4\pi r^2 = 4 \times 3 \cdot 1416 \times 3960^2 \text{ square miles.}$$

I have set out the calculations as straightforward multiplication, which anyone can verify, and the result is 197,061,258 square miles. But the end figures are certainly not reliable, because the earth is only roughly a sphere of 3960 miles radius. The most we can say from this calculation is that the area of the world is not far different from 197 million square miles.

$$
\begin{array}{r}
3960 \\
3960 \\
\hline
11880000 \\
3564000 \\
237600 \\
\hline
15681600 \\
4 \\
\hline
62726400 \\
3 \cdot 1416 \\
\hline
188179200 \\
6272640 \\
2509056 \\
62726 \\
37636 \\
\hline
197061258 \\
\end{array}
$$

There is a formula which gives the area of a spheroid just as exactly as we care to work it out. The formula is:

$$2\pi a^2 + \frac{\pi b^2}{e} \log \frac{1+e}{1-e}.$$

It is a tricky formula, and one has to be careful in using it. *a* and *b* are the equatorial and polar radii. $a = 3963 \cdot 34$ miles, and $b = 3949 \cdot 99$ miles. *e* is the eccentricity of the ellipse of a meridian; so $e = \frac{\sqrt{a^2 - b^2}}{a}$. The logarithm is the Napierian logarithm; we can find the common logarithm, and multiply it by $2 \cdot 30258$.

Here is the working:

$$2\pi a^2 = 2 \times 3 \cdot 1416 \times 3963 \cdot 34^2$$
$$= 98,697,000 \text{ square miles;}$$

$$e = \frac{\sqrt{a^2 - b^2}}{a} = \frac{\sqrt{3963 \cdot 34^2 - 3949 \cdot 99^2}}{3963 \cdot 34}$$
$$= \cdot 0820085,$$

$$\log \frac{1+e}{1-e} = \log_e \frac{1 \cdot 0820085}{\cdot 9179915} = \log_e 1 \cdot 178669$$
$$= \cdot 071392 \times 2 \cdot 30258$$
$$= \cdot 164386,$$

$$\frac{\pi b^2}{e} \log \frac{1+e}{1-e} = \frac{3 \cdot 1416 \times 3949 \cdot 99^2}{\cdot 0820085} \times \cdot 164386$$
$$= 98,253,000 \text{ square miles.}$$

The total area is:

$$\begin{array}{r} 98,697,000 \\ 98,253,000 \\ \hline 196,950,000 \text{ square miles.} \end{array}$$

So that 197 million square miles is a close enough approximation.

I have worked that out in order to draw attention to one or two points. The area thus found is the area of an exact spheroid. No allowance has been made for ups and downs, so that the area of the land and water surfaces is certainly greater than 197 million square miles.

*Whitaker* gives the estimated area of the earth as 196,550,000 square miles, but there is no explanation of what this estimate means. The five-figure accuracy is at least dubious.

*Physical and Chemical Constants* gives the area as $5 \cdot 12 \times 10^{18}$ square centimetres (land $1 \cdot 45 \times 10^{18}$ + water $3 \cdot 67 \times 10^{18}$), or rather more than five trillion square centimetres. It is always interesting to check one estimate against another by turning them both to the same unit. We know that:

$$1 \text{ square centimetre} = \cdot 1550 \text{ square inch.}$$

So:                 $5 \cdot 12 \times 10^{18}$ square centimetres

$$= 5 \cdot 12 \times \cdot 1550 \times 10^{18} \text{ square inches}$$

$$= \frac{5 \cdot 12 \times \cdot 1550 \times 10^{18}}{12^2 \times 5280^2} \text{ square miles.}$$

(We divide by $12^2$ to change square inches to square feet, and by $5280^2$ to change square feet to square miles.)

$$= 197 \cdot 7 \text{ million square miles.}$$

Again there is nothing to indicate what exactly is meant by the area. It is evidently not the area of the exact spheroid.

Geographers should certainly make allowance for the inequalities of land surface; indeed, they do seem to try to make such an allowance. I have an atlas which gives for the area of the earth the extraordinary quantity:

$$195,647,240 \text{ square miles.}$$

The number is extraordinary because it claims at least 8-figure accuracy, and that in itself is sufficient reason for rejecting it. We can readily find the radius of the sphere of which this is the area.

$$4\pi r^2 = 195,647,240 \text{ square miles,}$$

$$r^2 = \frac{195,647,240}{4\pi} \text{ square miles,}$$

and $r$ is the square root.

$$r = 3946 \text{ miles.}$$

The radii of the earth given in the atlas are 3950 miles and 3963 miles, both greater than 3946 miles. So there is some discrepancy, in spite of the nice air of precision about the stated area. In hunting for the origin of the extraordinary area, I found in the atlas a list of the areas of the oceans and continents. These included 17,074,080 square miles for Asia, and a vague 2,500,000 square miles for Antarctica. Someone had added these strangely different numbers without, apparently, any conception of what they meant.

Out of curiosity I turned up another atlas which gave the areas of the oceans and continents. A comparison of the two, to which

I have added *Whitaker's* numbers, certainly turned out to be curious. Here it is:

|  | Atlas A Area in square miles | Atlas B Area in square miles | Whitaker Area in square miles |
|---|---|---|---|
| Europe | 3,864,740 | 3,900,000 | 3,750,000 |
| Asia | 17,074,050 | 16,500,000 | 17,000,000 |
| Africa | 11,521,530 | 11,500,000 | 11,500,000 |
| Australasia | 3,450,220 | 3,600,000 (Oceania) | 3,450,000 |
| North America | 9,294,330 | 8,700,000 | 8,000,000 |
| South America | 6,817,390 | 7,300,000 | 6,800,000 |
| Antarctica | 2,500,000 | 5,000,000 | 5,000,000 |
| Total land area | 54,522,260 | 56,500,000 | 55,500,000 |

It is much easier to find the volume of a spheroid than to find its area, and here at least, one would have thought, there need be no discrepancies. We can get a good estimate of the volume of the earth by regarding it as a sphere of radius 3960 miles. The volume is given by: $\frac{4}{3}\pi r^3$.

We have already found that:

$$4\pi r^2 = 197 \cdot 06 \times 10^6 \text{ square miles.}$$

So we need only multiply this by $\dfrac{r}{3} = \dfrac{3960}{3} = 1320$.

$$197 \cdot 06 \times 1320 \times 10^6 \text{ cubic miles}$$
$$= 260,000 \times 10^6 \text{ cubic miles}$$
$$= \cdot 26 \times 10^{12} \text{ cubic miles.}$$

That is, the volume of the earth is rather more than a quarter of a billion cubic miles.

We should get a more accurate result for the volume of the earth by using the measured and calculated radii: the two equal equatorial radii, and the polar radius. Instead of $r^3$ we have the product of the radii, $a^2b$.

$$\tfrac{4}{3}\pi a^2 b = \tfrac{4}{3} \times 3 \cdot 1416 \times (3963\tfrac{1}{3})^2 \times 3950$$
$$= \cdot 2599 \times 10^{12} \text{ cubic miles.}$$

This more exact result shows that ·26 billion cubic miles is a very good approximation.

Turning to the book of *Constants* we find the volume of the earth given as:

$$1\cdot085 \times 10^{27} \text{ cubic centimetres.}$$

For comparison, we will turn this into cubic miles.

$$1 \text{ cubic centimetre} = 0\cdot0610 \text{ cubic inch.}$$

So:

$$1\cdot085 \times 10^{27} \text{ cubic centimetres}$$
$$= 1\cdot085 \times \cdot0610 \times 10^{27} \text{ cubic inches}$$
$$= \frac{1\cdot085 \times \cdot0610 \times 10^{27}}{12^3 \times 5280^3} \text{ cubic miles.}$$

(We divide by $12^3$ to change cubic inches to cubic feet, and by $5280^3$ to change cubic feet to cubic miles.)

$$= \cdot2602 \times 10^{12} \text{ cubic miles.}$$

This seems to be rather too much, especially as the radii obtained by the U.S. Survey, which are here used, are greater than those obtained by other measurements.

The atlases both give the volume of the earth as 259,880,000,000 cubic miles. I think they should have called it 260,000,000,000 cubic miles.

The volume we can give with such accuracy, assuming the correctness of the measurements of the radii, is the volume that would exist if all the continents were pared down to the level of the sea. Something ought to be added for the parts which stand up above that level; but it is not easy to find out how much. Anyone might be excused for giving up the attempt in despair; but it is possible to form some sort of estimate.

We need large-scale contour maps of the continents and they should be maps in which equal areas of the map represent equal areas of the country. We can get the measurements by using squared paper.

We can use the squared paper to estimate the area of a whole continent. Then we take it in sections; we find the area between each contour and the next contour above it. We have a useful check on the work, because the sum of the areas between the contours should equal the whole area of the continent.

Suppose we have measured the area between the 4000-foot and 5000-foot contours. We take the average height of this area as

4500 feet; that is probably the best we can do. 4500 feet $= \frac{4500}{5280}$ miles. The volume of this piece of land above sea-level is, in cubic miles:

$$\text{measured area in square miles} \times \tfrac{4500}{5280}.$$

We take the spaces between the other contours, and deal with them in the same way. Addition gives the total volume of land above sea-level for the whole continent. Division of this total volume by the area of the continent gives the average height in miles. Multiplication by 5280 gives the height in feet. We can of course omit 5280 altogether, because it comes into both numerator and denominator. The working is:

$$\frac{\substack{\text{sum of areas between contours} \\ \text{in square miles}} \times \substack{\text{their average heights} \\ \text{in feet}}}{\text{total area in square miles}}.$$

The continent with the greatest average height is Asia, as one might expect. I have two estimates of the average height; one is 3120 feet, and the other 3189 feet. We have seen that the measurement is one of considerable difficulty, and the approximations we have to make are rather drastic. Add to this the vagueness of geographical knowledge about Central Asia, and it seems just a little absurd to pretend to have estimated the average height to within a foot, as 3189 feet does. We can get a better approximation. $\frac{3}{5}$ of a mile is 3168 feet, and as this happens to be about the average of the two estimates, we can take the average height as $\frac{3}{5}$ mile. $\frac{3}{5}$ mile is a vague height, purposely vague because the estimate is itself vague and is most correctly expressed in a vague way.

And now look at the average heights for the other continents; they are taken from the same two sources:

*Africa*: 2021 feet and 2130 feet. The difference is about 5 per cent; not too bad, but the end figures seem to be pretence.

*North America*: 1888 feet and 2300 feet. The difference is about 22 per cent; not too good.

*South America*: 2078 feet and 1970 feet. The difference is about 5 per cent; again not too bad.

*Australia*: 805 feet and 1310 feet. There is no reconciling these two estimates. And how much exact knowledge have we of central Australia?

*Europe*: 939 feet and 930 feet. This is pretty fair agreement, and that is what we should expect. We have far more detailed information about the contours of Europe than about those of the other continents.

It looks as if the difficulty of finding the average height of a continent had been underestimated. The most I would say with the available information is that the average heights of the continents are: Asia—$\frac{3}{5}$ mile; North and South America, and Africa—each about $\frac{2}{5}$ mile; Australia and Europe—each about $\frac{1}{5}$ mile.

And now let us look at the masses of land above sea-level in the continents.

*Asia*: The area is about 17 million square miles. $17 \times \frac{3}{5} = 10 \cdot 2$. So we can take the volume of land above sea-level as about 10 million cubic miles.

*Africa*: Area $11\frac{1}{2}$ million square miles. $11\frac{1}{2} \times \frac{2}{5} = 4\frac{3}{5}$. So the volume above sea-level is about $4\frac{1}{2}$ million cubic miles.

*North America*: Area $9\frac{1}{4}$ million square miles. $9\frac{1}{4} \times \frac{2}{5} = 3 \cdot 7$. So the volume above sea-level is about $3\frac{3}{4}$ million cubic miles.

*South America*: Area $6\frac{3}{4}$ million square miles. $6\frac{3}{4} \times \frac{2}{5} = 2 \cdot 7$. So the volume above sea-level is about $2\frac{3}{4}$ million cubic miles.

*Australasia*: Area $3\frac{1}{2}$ million square miles. $3\frac{1}{2} \times \frac{1}{5} = \cdot 7$. So the volume above sea-level is about $\frac{3}{4}$ million cubic miles.

*Europe*: Area $3\frac{3}{4}$ million square miles. $3\frac{3}{4} \times \frac{1}{5} = \frac{3}{4}$. So the volume above sea-level is about $\frac{3}{4}$ million cubic miles. Perhaps this is too high. If we take 930 feet as a fairly accurate estimate of the average height the volume is:

$$3\frac{3}{4} \times \tfrac{930}{5280} = \cdot 66.$$

This gives a volume of $\frac{2}{3}$ million cubic miles.

The total volume above sea-level is therefore about:

$$10 + 4\frac{1}{2} + 3\frac{3}{4} + 2\frac{3}{4} + \frac{3}{4} + \frac{2}{3} = \text{about } 22\frac{1}{2} \text{ million cubic miles.}$$

If we compare that with the whole volume of the earth, $\cdot 26$ billion cubic miles, or 260,000 million cubic miles, we see how insignificant it is, so that no correction is necessary in the whole volume of the earth. The actual fraction is $\dfrac{22\frac{1}{2}}{260{,}000}$, or rather less than one ten-thousandth.

The volumes we have found for the continents show how much Asia stands out above the rest. Its volume is about $\frac{4}{9}$ of the total volume of all the continents.

It may give a clearer picture if we call the volume of Asia 100, and so get rid of most of the fractions. The comparison is:

| Asia | 100 | Africa | 45 |
| N. America | 37 | S. America | 27 |
| Australasia | 7 | Europe | $6\frac{1}{2}$ |

Rather more than two-thirds of the whole is contained in the Asia-Africa-Europe group, and less than one-third in the Americas.

The insignificance of the continental heights in adding to the radius of the world may be seen if we imagine the continental volumes spread evenly over the world. We can take the area of the earth as roughly 200 million square miles, and the continental volumes as a little over 20 million cubic miles. The average additional height would be $\frac{20}{200} = \frac{1}{10}$ mile.

I have before me an estimate of the 'total sea area', which is recklessly given as:

$$141,124,980 \text{ square miles.}$$

The mean depth of the sea is given with almost equal recklessness as 11,470 feet. One at once begins wondering how much we actually know about the configuration of the sea floor, and how it is possible to form so close an estimate of areas which include frozen seas, where the difference between land and water is not easily distinguished, and vaguely explored coasts.

We might reasonably take the area of the oceans as about 140 million square miles, and the average depth as about 2 miles. There seems to be no doubt that the average depth of the oceans is considerably greater than the average height of the continents. It is somewhere about five times as great.

The whole volume of water in the oceans is something like $140 \times 2 = 280$ million cubic miles; that is, the area in square miles, multiplied by the average depth in miles. There is enough water in the oceans to cover the 200 million square miles of surface to a depth of a mile and a half or so. If, and when, erosion is finally complete, and the continents have been shaved off down to their bases and spread evenly over the sea floor, the final state of the earth should be a solid globe with more than a mile deep of water covering it. How long the uneasy crumpling of the earth's crust will postpone that doom it would be a hard task to estimate.

The water of the oceans is so vast an amount that it forms a considerable part of the volume of the whole world. The fraction is: $\frac{280}{260000}$ = rather more than $\frac{1}{1000}$, or more than one part in a thousand. And it is something like thirteen times the volume of the whole mass of land above sea-level.

About $1\frac{1}{4}$ million gallons of water pours over Niagara in a second, and as there are about 31 million seconds in a year, that is about 40 billion gallons in a year. How long would it take the water of the oceans to pour over Niagara?

$$280 \times 10^6 \text{ cubic miles}$$
$$= 280 \times 5280^3 \times 10^6 \text{ cubic feet}$$
$$= 280 \times 5280^3 \times 6\frac{1}{4} \times 10^6 \text{ gallons}$$
$$= 2{\cdot}575 \times 10^{20} \text{ gallons}$$
$$= 257\frac{1}{2} \text{ trillion gallons.}$$

So the length of time to pour over Niagara:

$$= \frac{257\frac{1}{2} \times 10^{18}}{40 \times 10^{12}} \text{ years}$$
$$= \text{about 6 million years.}$$

Water is so excellent a solvent, and there have been so many opportunities for solution, that there are few substances that are not dissolved in sea water even if only in all but infinitesimal amounts. I once read a story about a man who invented an apparatus for collecting gold from the sea. The apparatus was fixed on the coast, and sea water ran through it and gave up its gold as it went. Alas for such dreams! The inventor seems to have had no conception of the time it would take for a cubic mile of water to run through his apparatus and deposit (*his* word) its few grains of gold. The apparatus seems to have dealt with 10,000 gallons of water in a minute—'some going!' as the inventor's friend exclaimed. Well:

$$1 \text{ cubic mile} = 5280^3 \text{ cubic feet}$$
$$= 5280^3 \times 6\frac{1}{4} \text{ gallons.}$$

So a cubic mile would take:

$$\frac{5280^3 \times 6\frac{1}{4}}{10,000} \text{ minutes} = \frac{5280^3 \times 6\frac{1}{4}}{10,000 \times 60 \times 24 \times 365\frac{1}{4}} \text{ years.}$$

$$= 175 \text{ years.}$$

It seems a long time to have to wait for a few grains of gold.

Various estimates have been made of the amount of radium in sea water. There appears to be something like one part in a thousand billion. In a ton of sea water there is:

$$\frac{2240 \times 16}{10^{15}} \text{ ounce of radium}$$

$$= 3\cdot6 \times 10^{-11} \text{ ounce per ton,}$$

or about $3\frac{1}{2}$ ounces in 100,000 million tons.

Only the ferocity of radioactivity makes it possible to detect so small an amount.

Now let us find the weight of the sea.

1 cubic foot of sea water weighs about 64 pounds.

1 cubic mile weighs $64 \times 5280^3$ pounds.

280 million cubic miles weigh $64 \times 5280^3 \times 280 \times 10^6$ pounds

$$= \frac{64 \times 5280^3 \times 280 \times 10^6}{2240} \text{ tons}$$

$$= 1\cdot2 \text{ trillion tons.}$$

The weight of radium in this, one part in $10^{15}$, is:

$$\frac{1\cdot2 \times 10^{18}}{10^{15}} = 1200 \text{ tons.}$$

The almost infinitesimal amount of $3\frac{1}{2}$ ounces in a hundred thousand million tons, mounts up to 1200 tons in all the oceans.

Before leaving the oceans we ought to look at a problem that has been a special trial to geographers. It is a pleasant idea to think of the Gulf Stream flowing north from the equator, swinging across the Atlantic, and bringing amelioration of winter conditions to an otherwise frozen coastline. It seems a lot to expect from one comparatively small current. Certainly the Gulf Stream leaves the Florida Channel with a high temperature, about 80° F. In summer it flows north to the latitude of Nova Scotia with little loss of temperature; but in winter the temperature falls to about 60° F., and winter is the season we have to worry about. But that is not all; the Gulf Stream meets the Labrador Current, with a stream of icebergs, in early summer, and this still further reduces its temperature till it is little above that of the surrounding sea. Even then, if it is going to do any good to northern climates, the Gulf Stream should be moving toward the north-east. Actually it turns vaguely eastward. It looks as if the Gulf Stream is a wash-out, so far as any considerable influence on the climate of western Europe is concerned. The most that it can do is to bring into northern waters a stream of water travelling at perhaps 20 miles in a day, 60 miles wide, and 600 feet deep; the temperature of this water would be a little above that of surrounding seas. The total volume of water passing a fixed line is:

$$20 \text{ miles} \times 60 \text{ miles} \times 600 \text{ feet} = 20 \times 60 \times \tfrac{600}{5280} \text{ cubic miles per day,}$$
$$= \text{nearly } 140 \text{ cubic miles per day.}$$

In the 6 months of the winter half of the year the total amount of warmish water that would flow in is:

$$180 \times 140 = \text{about } 25,000 \text{ cubic miles.}$$

That may seem a considerable amount, until we compare it with the whole mass of water in the North Atlantic, which is about 25 million cubic miles. So if we accept the Gulf Stream theory we have to believe that less than a thousandth of the ocean and probably considerably less, a few degrees above the general level of temperature, can affect climate so greatly as to make the difference between frozen Labrador and the ice-free ports of Norway.

The alternative theory is the drift north-eastward of the upper half of the North Atlantic. The temperature of the lower half of the North Atlantic is down at freezing point, or even lower, whilst the upper half is at 40° or over. Thus there is a continual interchange of Arctic water flowing toward the equator and a surface outflow of tropical water. A drift of a mile a day would bring 2000 cubic miles per day (2000 miles wide and a mile deep) of warmish water toward the north-east; and in 180 days that is 360 thousand cubic miles, as compared with a highly problematic 25 thousand cubic miles from the Gulf Stream.

The drift is toward the north-east because the water starts from the tropics with a greater eastward velocity than the ocean floor it is drifting over. Whereas Arctic water is moving toward regions with a greater eastward velocity, and is therefore piled up toward the west. It is a confirmation of this theory that along the equator there is cold Arctic and Antarctic water at a comparatively shallow depth.

# CHAPTER 3

## *The Arithmetic of Great Rivers*

RECURRING DISASTERS have drawn pained attention to the arithmetic of great rivers and their basins.

The greatest of all rivers is the Amazon. The Missouri-Mississippi is longer, but the Amazon has the greatest basin, and carries the greatest volume of water to the sea. The area of the basin has been estimated at $2\frac{1}{2}$ million square miles, and that is two-thirds the area of Europe.

The Amazon basin is in the region of heavy equatorial rain; this is increased by the north-east trade winds which blow in from the sea, and by the mountains which rob the winds of most of their moisture. In the dampest parts the rainfall is more than 80 inches a year, and it appears to be more than 5 feet per annum over the whole basin. The volume of water which falls in the Amazon basin yearly is therefore:

$$2,500,000 \times \tfrac{5}{5280} \text{ cubic miles}$$
$$= \text{nearly } 2400 \text{ cubic miles.}$$

That is an almost incredible amount of water to fall on land, and it sets the river an almost insoluble problem: what to do with it?

In many great rivers the fraction of the total rainfall over the whole basin which flows down the river to the sea is about a quarter. In a damp region like the Amazon basin the proportion is probably higher. To be on the side of understatement rather than overstatement, we can take the fraction as a quarter. That still leaves 600 cubic miles of water to be carried down to the sea.

A cubic mile is an uncomfortable measure for water, unless it happens to be water in the sea; the imagination boggles at the idea of a mass of water a mile high. We could imagine it spread evenly over the whole of England, which has an area of about 50,000 square miles. Over this area the depth would be:

$$\frac{2,500,000}{50,000} \times 5 \text{ feet} = 250 \text{ feet.}$$

In gallons the measure is:

$$600 \text{ cubic miles} = 600 \times 5280^3 \text{ cubic feet}$$
$$= 600 \times 5280^3 \times 6\tfrac{1}{4} \text{ gallons}$$
$$= 550 \text{ billion gallons.}$$

That is the number of gallons that flow down the Amazon in a year. The daily flow is:

$$\tfrac{550}{365} \text{ billion gallons} = \text{about } 1\tfrac{1}{2} \text{ billion gallons.}$$

We will also find the flow per second. This is:

$$\frac{1\tfrac{1}{2} \times 10^{12}}{24 \times 3600} = \text{more than } 17 \text{ million gallons per second.}$$

To compare big things with small, we can compare the Amazon with the Thames. The Thames hardly counts among great rivers; it is sometimes tucked in at the end in English atlases; but its claims to greatness have little to do with size. During a year of heavy rainfall the flow of the Thames has risen to 18,000 gallons per second. That affords a striking comparison with the Amazon. The flow of the flooded Thames is just about the thousandth part of the usual flow of the Amazon. A thousand rivers like the Thames would barely suffice to carry off the flood of water that passes along the Amazon.

And here is another comparison from the same spot. London uses 300 million gallons of water daily, and that is about 40 gallons for each inhabitant. We have seen that the flow of the Amazon is 17 million gallons per second.

$$\tfrac{300}{17} = 17 \cdot 6 \text{ or } 17\tfrac{1}{2}.$$

So that about $17\tfrac{1}{2}$ seconds' flow of the Amazon would suffice to supply London with water for 24 hours. The whole flow of the Amazon would supply:

$$\frac{1\tfrac{1}{2} \text{ billion gallons per day}}{300 \text{ million gallons per day}}$$

$= 5000$ cities the size of London with a perennial water supply. That is of course far more than the population of the world, which is estimated at something over 2000 millions.

$$5000 \times 8 \text{ millions} = 40,000 \text{ millions,}$$
$$\frac{40,000}{2000} = 20.$$

So that the Amazon would supply everyone in the world with $20 \times 40 = 800$ gallons per day, or 20 times as much for each inhabitant of the world as is used by the average London citizen.

The 600 cubic miles of water that flows yearly along the Amazon is so vast an amount as to be altogether beyond control. In the rainy season the Amazon is more like a lake than a river; an elongated lake stretching across a continent. It is said to be more than 400 miles wide in places, and even the narrows are five or six miles wide. Those who live near the Amazon have had to adapt their lives to the vagaries of a river subject to widespread floods that in other conditions would be disastrous. This adaptation replaces attempts to control the river. One can hardly imagine the mighty Amazon being tamed by the devices of river engineers, and flowing gently between embankments.

The Mississippi presents a different problem, and it is a real problem for river engineers. The fertile land along the river, fertilised by the river itself, is highly cultivated, and produces abundant crops; and it has a large population. In these conditions widespread floods might easily be disastrous, and sometimes are. Some sort of control of the river is not merely desirable, but a necessity of life in the Mississippi valley.

The basin of the great Missouri-Mississippi system has an area of about $1\frac{1}{4}$ million square miles; that is, it has half the area of the Amazon basin. The annual rainfall also is about half, that is, it is 30 inches. So that the total amount of water that falls on the Mississippi basin is a quarter of what falls on the Amazon basin. The proportion of the rainfall that flows down the river is known to be about a quarter; this is the fraction we took as a minimum for the Amazon. Thus the average discharge of the Mississippi is a quarter of the 17 million gallons per second we found for the Amazon. That is, it is:

$$4\frac{1}{4} \text{ million gallons per second.}$$

If we happen to want this quantity in cubic feet, it is:

$$\frac{4\frac{1}{4}}{6\frac{1}{4}} \text{ million cubic feet per second}$$

$$= \cdot 68 \text{ million cubic feet per second,}$$

or over two-thirds of a million cubic feet per second.

The size of the Mississippi near its mouth has been carefully measured; it is the cross-section area we want. To find this we have to measure the depth of the river at regular intervals. The measured depths give the contour of the river bed; we can use them to find the cross-section area. This area is 200,000 square feet.

We can find the average speed of the river throughout the year in this way: we imagine the water which flows through in a second to be behind the measured cross-section. The volume of this water is ·68 million, or 680,000 cubic feet. So it stretches back a distance of:

$$\frac{680,000}{200,000} = 3·4 \text{ feet.}$$

Now the whole of this water flows through in a second, so the rate of flow is 3·4 feet per second. We want to change this speed into miles per hour.

$$3·4 \text{ feet per second} = 3·4 \times 3600 \text{ feet per hour}$$
$$= \frac{3·4 \times 3600}{5280} \text{ miles per hour}$$
$$= 2·3 \text{ miles per hour.}$$

That is a reasonable rate of flow for a great river flowing over almost level land. We could check the speed by measuring it at different points across the width of the river. The speed we have found is of course sufficient to carry off the normal flow of the river; otherwise the river would widen its banks, or scour its bed still deeper; it would do something about it. Trouble arises when there is exceptionally heavy rainfall, especially when heavy rain continues over a prolonged wet period. The average rainfall in the Mississippi basin, we say, is 30 inches per annum. That is $\frac{30}{385}$ inch per day, or just about a twelfth of an inch per day. In periods of heavy rainfall there may be an inch or more of rain in a day over a large part of the basin, perhaps over the basin of one of the great tributaries. Much more than the usual quarter goes to the river, because evaporation has stopped. Normally a lot of rain is absorbed by the soil, and returned slowly to the river and the sea; during heavy rain this easing of the situation ceases, or is greatly reduced. Swollen tributaries pour ten times the usual amount of water, or more, into the Mississippi valley. The river cannot carry away so vast an influx of water, even though the higher level of the river in the upper, flooded, part of the valley increases the slope, and so the rate of flow. The water pours out over the banks, and the valley is flooded.

By comparison with the Amazon and the Mississippi, the St Lawrence is a most fortunate and well-behaved river. Variations in the water level are small, and floods are a minor feature. The reason is that the Great Lakes provide a great storage reservoir for flood water. The drainage system reaches an obvious bottleneck at the Niagara Falls; there is usually this sort of bottleneck in a drainage system, but it is not always so obvious. The drainage area above the Niagara Falls is estimated at a quarter of a million square miles. The surplus water from the whole of this area pours over the Falls. Various estimates have been made of the flow of water over the Falls. Older estimates placed it at 400,000 cubic feet per second. Recent estimates have reduced this to the more probable rate of about 200,000 cubic feet per second, or $1\frac{1}{4}$ million gallons per second.

It is easy to show that the newer estimate is more probable. The rainfall is in the region of 30 inches per annum, or $2\frac{1}{2}$ feet per annum, so that the total volume of water is:

$$250,000 \times 5280^2 \times 2\frac{1}{2} \text{ cubic feet per annum}$$
$$= \frac{250,000 \times 5280^2 \times 2\frac{1}{2}}{31 \times 10^6} \text{ cubic feet per second}$$
$$= 560,000 \text{ cubic feet per second.}$$

200,000 cubic feet per second is rather more than a third of this, and that is more probable than two-thirds.

The drainage area includes about 95,000 square miles of the Great Lakes. During the year the lakes fluctuate in height, rising higher in wet weather, and falling during dry periods. The fluctuations are not very great; they amount to from $2\frac{1}{2}$ feet to 3 feet during the year. Certainly not a great range of fluctuation; but we have to remember that this variation in height extends over the whole 95,000 square miles of lake surface.

The area of 95,000 square miles $= 95,000 \times 5280^2$ square feet. So the volume of a space with this area, and $2\frac{1}{2}$ feet high, is:

$$95,000 \times 5280^2 \times 2\frac{1}{2} \text{ cubic feet}$$
$$= 6 \cdot 6 \text{ billion cubic feet.}$$

That is the vast volume of water that can be retained by a rise of $2\frac{1}{2}$ feet in the level of the Lakes.

We can see how long this additional volume of water would take to pour over Niagara. The average rate of flow is about 200,000 cubic feet per second. 6·6 billion cubic feet would pour over in:

$$\frac{6\cdot6 \times 10^{12}}{200,000} \text{ seconds}$$

$= 33$ million seconds, or rather more than a year.

Thus a rise of $2\frac{1}{2}$ feet in the level of the lakes can hold up a whole year's normal flow. The additional water is carried off gradually by an increase in the rate of flow up to 270,000 cubic feet per second, or even more.

A river with as great a basin as that of the St Lawrence is usually subject to flooding; the St Lawrence is exceptional in having the Great Lakes as a safety valve. The freedom from flooding has suggested a means of preventing, or at least mitigating, floods in other rivers. It has been proposed to create large temporary artificial lakes, as a means of storing flood waters in times of heavy rain, so that the river may have a longer period in which to carry off the flood water. The flood areas would be long and narrow; they would be confined by low dykes of earth; they would have few inhabitants. In times of stress, when swollen rivers were beginning to overflow their banks, the few inhabitants of the flood areas would be evacuated, and water would then be allowed to enter through the dykes, so as to reduce the amount of water in the river. In the Fen District of eastern England, long narrow areas have been reserved for flooding in this way; they have purposely been made narrow so that they can be evacuated rapidly in time of need. Large-scale flood areas have been proposed as a means of mitigating the effects of floods on the Mississippi. One of the difficulties standing in the way of this scheme is that the fertile land of the valley is extremely valuable.

The alternative scheme of confining great rivers between embankments has not been too successful. The Thames Embankment has usually preserved London from flooding, but not the

Thames valley. The levees, or embankments, of the lower Mississippi have saved New Orleans, but they have caused grave anxiety from time to time. China's river problem is the Hwangho, 'the sorrow of China'. Embankments have been built to restrain the river, but the bed has silted up, and the embankments had to be raised still higher; so that the river bed is high above the surrounding land, and floods bring widespread disaster when the embankments are breached. It should be clear that the problems facing river engineers are extremely difficult, especially when the engineers are dealing with really great rivers.

It is rather curious that most of the great rivers of the world should flow into the Atlantic. There are the two great equatorial rivers, the Amazon and the Congo; the Mississippi and the St Lawrence; and the Nile. Against these, the much bigger Pacific has only the two great Chinese rivers, the Hwangho and the Yangtse; and the Indian Ocean the Ganges and Irrawadi. It is of course the equatorial rivers that give the Atlantic the preponderance. The Amazon probably contains more fresh water than all the other rivers and lakes together. The Congo and the Niger drain the region of heavy equatorial summer rains of Africa, corresponding to the Amazon basin.

To return to the St Lawrence and the Niagara Falls, there is an interesting comparison between the flow over the Falls and the size of Windermere, which is the largest of the English lakes. Windermere is a small thing when compared with the Great Lakes, but it contains nevertheless a great volume of water. A fair estimate of the area of Windermere is 5·7 square miles, and it has the considerable average depth of 78½ feet. The average depth is the mean of a large number of soundings taken at different parts of the lake.

The volume of water in the lake is:

$$5\text{·}7 \text{ square miles} \times 78\tfrac{1}{2} \text{ feet}$$
$$= 5\text{·}7 \times 5280^2 \times 78\tfrac{1}{2} \text{ cubic feet}$$
$$= 1\text{·}247 \times 10^{10} \text{ cubic feet.}$$

This is over 12,000 million cubic feet

$$= 12{,}000 \times 6\tfrac{1}{4} \text{ million gallons}$$
$$= 75{,}000 \text{ million gallons.}$$

London uses 300 million gallons of water daily, so the water in Windermere would supply London for:

$$\frac{75,000}{300} = 250 \text{ days, or } 8\tfrac{1}{3} \text{ months.}$$

Now let us compare this great volume of water with the flow over Niagara. The flow over Niagara is 200,000 cubic feet per second, so that $1 \cdot 247 \times 10^{10}$ cubic feet would flow over the Falls in:

$$\frac{1 \cdot 247 \times 10^{10}}{2 \times 10^5} \text{ seconds}$$

$$= \cdot 624 \times 10^5 \text{ seconds}$$

$$= 62400 \text{ seconds}$$

$$= 1040 \text{ minutes}$$

$$= 17\tfrac{1}{3} \text{ hours.}$$

So that if Windermere, somehow or other, were to be emptied, the flow of water over Niagara Falls would suffice to fill it again in about 17 hours.

The greatest lake in the world is Lake Superior. Its length is given as 390 miles, its average width as 80 miles, and its average depth as 900 feet. In cubic miles the volume of water in Lake Superior is:

$$390 \times 80 \times \tfrac{900}{5280}$$

$$= 5320 \text{ cubic miles,}$$

which is just about twice the amount of rain that falls on the Amazon basin in a year.

To flow over Niagara this amount of water would take:

$$\frac{5320 \times 5280^3}{200,000} \text{ seconds}$$

$$= \frac{5320 \times 5280^3}{200,000 \times 3600 \times 24 \times 365\tfrac{1}{4}} \text{ years}$$

$$= 124 \text{ years.}$$

So that the water in Lake Superior would maintain the flow over Niagara for 124 years, compared with the 17 hours that Windermere would keep it going.

The tumbling waters of Niagara Falls have always attracted engineers as a source of power. Kipling's Explorer watched unharnessed rapids wasting fifty thousand head an hour (though I do not think 'an hour' should be there). If he had stared at Niagara he would have seen more than that.

The height of the Falls is about 160 feet and 200,000 cubic feet of water, falling that distance, would develop an enormous amount of energy. 200,000 cubic feet of water weighs $200,000 \times 62\frac{1}{2}$ pounds. The amount of energy developed in falling 160 feet is:

$$200,000 \times 62\frac{1}{2} \times 160 \text{ foot-pounds per second.}$$

Now    1 horse-power $= 33,000$ foot-pounds per minute
$$= 550 \text{ foot-pounds per second.}$$

So the total horse-power developed by the Falls is:

$$\frac{200,000 \times 62\frac{1}{2} \times 160}{550}$$
$$= \text{about } 3\frac{1}{2} \text{ million horse-power.}$$

In order to use the power of $3\frac{1}{2}$ million horses, available night and day at Niagara, we should have to harness the whole of the falling water, and deliver it to turbines placed at the foot of the Falls, so as to waste nothing.

The power needed for a large aeroplane is in the region of 4000 horse-power. So the Niagara Falls could supply energy enough to fly nearly 1000 of the biggest aeroplanes continuously, and that is not so many as one might have expected.

There is another problem arising out of the arithmetic of great rivers, that is not so urgent and personal as the problem of flooding. In this problem the sea plays a part as well as rivers; it is the problem of erosion. Occasionally there is a spectacular piece of erosion by the sea, when whole cliffs fall at once. It was a piece of erosion of this kind that brought tragedy to Norwegian sea villages a few years ago. But that kind of erosion is exceptional. Year in, year out, the sea brings down masses of rock without injury to anyone; and at the same time erosion by rivers and rain goes remorselessly on at equal speed.

The waters of rivers often seem clear enough, and erosion by rivers seems so small as to be negligible. But there is always some solid matter being carried along by rivers, and in times of flood, when river water is really muddy, the amount of river erosion may be considerable. The Mississippi has been slangily and affectionately called Big Muddy, because of the muddiness of its waters, and the amount of mud it takes down to the sea. Careful measurement of the water of the Mississippi shows that there is no less than one cubic foot of solid matter in every 3000 cubic feet. We have seen that the annual flow of the Mississippi is in the region of 150 cubic miles. Of this volume, one part in 3000 is solid matter; so the river carries down to the sea every year:

$$\frac{150}{3000} = \frac{1}{20} \text{ cubic mile of solid matter.}$$

That is a truly enormous amount of matter—enough to cover a space a mile square and 264 feet deep. We can perhaps appreciate better what it means if we find its weight in tons. We begin by changing the volume into cubic feet:

$$\frac{1}{20} \text{ cubic mile} = \frac{5280^3}{20} \text{ cubic feet}$$
$$= 7360 \text{ million cubic feet.}$$

If this volume of material were water, it would weigh $62\frac{1}{2}$ pounds per cubic foot. We can take the density as twice as great. This would give for the weight of the $\frac{1}{20}$ of a cubic mile:

$$\frac{7360 \times 62\frac{1}{2} \times 2 \times 10^6}{2240} \text{ tons}$$
$$= \text{over 400 million tons.}$$

That is the weight of fine solid matter that is carried down yearly by the Mississippi. Most of it is so fine that it is suspended in the running water. It is not till it reaches the still water of the sea that it settles to the bottom. Some of it is deposited in the delta which sprawls out at the mouth of the river, some is deposited on the sea floor, and some is carried out and deposited along surrounding coasts, and silts up lagoons.

This twentieth of a cubic mile is taken unevenly from the Mississippi basin, but we can imagine it spread out evenly, so that we can find the average rate of erosion. Spread out over so great an area we should not expect the effect to be very great in a single year; it is only the accumulation of the silt from a large area that appears great in the comparatively narrow waters of the river. We have to think of a twentieth of a cubic mile stripped evenly from a basin with an area of $1\frac{1}{4}$ million square miles. We divide $\frac{1}{20}$ cubic mile by this area, and find:

$$\frac{\frac{1}{20} \text{ cubic mile}}{1,250,000 \text{ square miles}} = \frac{1}{25,000,000} \text{ mile.}$$

(We could leave that as one mile in 25 million years, but it is more helpful to turn it into inches.)

$$= \frac{12 \times 5280}{25,000,000} \text{ inch}$$
$$= \cdot 0025 \text{ inch}.$$

That is, erosion in the Mississippi basin is at the rate of ·0025 inch yearly. We can say that the erosion is 25 inches in 10,000 years, or one inch in 400 years.

We have seen that the average height of the North American continent is given as 2300 feet. At the rate of erosion we have found for the Mississippi basin the whole continent would be reduced to sea-level in

$$2300 \times 12 \times 400 \text{ years}$$
$$= \text{about 11 million years}.$$

Thinking in the massive strain familiar to geologists, it seems a short span of life for a continent. However, we need not despair of North America. Much may happen in 11 million years. The rate of erosion will certainly slow down as the basin becomes flatter; and the forces that fold strata into mountains are still at work. There is every prospect of a much longer lease of life than a mere 11 million years or so. We can indeed have considerable assurance of the permanence of North America for a long time to come.

Let us think rather of the British Isles, and especially of England, so precariously exposed to sea erosion. It has been estimated that the coast of England, by sea erosion alone, is being washed away at an average rate of one foot per annum along the whole coast. That may not seem much; it would not alarm one greatly, even if one happened to notice it, to return to a seaside spot after a year's absence, and to find a foot of the shore gone. On the other hand, it may seem a very rapid rate of erosion to those who remember how such things accumulate over long periods of time. A foot in a year is 70 feet in the lifetime of the psalmist, and that is a considerable loss. And it is a mile in about five thousand years.

Let us look at the matter in another way; let us see what area of land is lost by sea erosion every year. It is not easy to estimate the length of a coastline, but we can take 1800 miles as a not unreasonable estimate of the length of the English coast. A foot depth along the whole of this coast would give an area of:

$$1800 \text{ miles} \times 1 \text{ foot}$$
$$= 1800 \times 1760 \times \tfrac{1}{3} \text{ square yards}$$
$$= \frac{1800 \times 1760}{3 \times 4840} \text{ acres}$$
$$= \text{nearly } 220 \text{ acres.}$$

That is the area that England is losing yearly by sea erosion: 220 acres, or a piece of land the size of a small farm.

And again we have to look ahead geologically, just as historically we learn to look back one thousand, two thousand or more years. In a thousand years the sea will have robbed us of:

$$220,000 \text{ acres}$$
$$= 220,000 \div 640 \text{ square miles}$$
$$= \text{about } 340 \text{ square miles.}$$

And that is quite a considerable tract of land. Our remote descendants may find themselves more and more constricted.

Now let us see what is being lost in volume and weight. Taking one part with another the average height of the cliffs round England is estimated at 50 feet. So the volume eroded is:

1 foot depth × 1800 miles long × 50 feet high.

(We want all the units alike, so we change the miles to feet.)

$$= 1800 \times 5280 \times 50 \text{ cubic feet}$$
$$= 475 \text{ million cubic feet,}$$

which is the volume of material eroded yearly by the sea.

Taking the same average density of twice that of water, the weight of this material is:

$$475 \times 10^6 \times 62\tfrac{1}{2} \times 2 \text{ pounds}$$
$$= \frac{475 \times 10^6 \times 62\tfrac{1}{2} \times 2}{2240} \text{ tons}$$
$$= \text{about } 26\tfrac{1}{2} \text{ million tons.}$$

That is just about an eighth part of the amount of coal burnt in England every year. So it seems that the sea is more merciful than man; man's erosion of coal is eight times more rapid than erosion by the sea.

The amount of erosion by rivers is estimated to be just about the same as erosion by the sea. That is to say, about 26½ million tons is swept off the face of England yearly, and deposited in the seas round the coasts. We can readily find the average depth represented by this amount of erosion. We have only to find:

$$\frac{\text{volume of erosion}}{\text{area of England}}$$

$$= \frac{475 \times 10^6 \text{ cubic feet}}{50,000 \text{ square miles}}$$

$$= \frac{475 \times 10^6}{50,000 \times 5280^2} \text{ feet}$$

$$= \frac{475 \times 10^6 \times 12}{50,000 \times 5280^2} \text{ inch}$$

$$= \cdot 004 \text{ inch per year.}$$

That is 4 inches in 1000 years, or 1 inch in 250 years.

The question arises as to which kind of erosion would wipe out England first. *Will* there always be an England? Coast erosion is rapid, even though it is only an eighth of coal erosion. If it proceeds unchecked at its present rate, the sea will have advanced 150 miles inland in

$$150 \times 5280 = 792,000 \text{ years.}$$

After A.D. 700,000 England would be a mere group of tough small islands surrounded by shoals and reefs, dangerous to shipping, if there is any shipping in those days to be dangerous to.

Whilst this is going on, river erosion is wearing down the land at the rate of an inch in 250 years. Allowing an average height of 500 feet, England would be reduced to the merest stump, by river and rain erosion, in:

$$500 \times 12 \times 250 = 1,500,000 \text{ years.}$$

It looks as if the waves are the more dangerous and the more immediately pressing enemies. But the prospect is not so alarming as it sounds. That is only one side of the picture. The tough core of England will slow down erosion by the sea; and as the land gets flatter the rivers will have less erosive effect. New land is being formed out of materials brought down by sea and rivers, especially about the Wash, and on parts of the south coast. And we can always hope that the erosion that moves mountains may be balanced by one of those upward heaves that lift them up again. Considering all things, or at least many things, there is no call for excessive geological pessimism regarding the remote future of England.

# CHAPTER 4

## *Widdershins*

THE FIRST of the celestial movements we are aware of is the movement of the sun across the sky from left to right. The movement is sometimes described as 'from east to west', but that is not a good description; it might equally well describe a movement in the opposite direction, through the north. 'From east to west through the south' is more accurate; 'from left to right' is simpler. The apparent movement of the sun is always from left to right, day and night alike. If anyone objects that by turning the back on the thing described we change over to 'from right to left', the answer is that common sense suggests facing the thing you are describing.

We have come to regard the direction left to right, or east to west through the south, as normal; and the reverse direction, right to left or east to west through the north, as abnormal. We have the word 'deasil' for the direction of the sun's daily course, and the better known word 'widdershins' for the contrary direction. Our clocks and watches turn deasil, and it would seem odd, almost contrary to nature, to have them turn widdershins. We read deasil, and Macaulay's widdershins writing 'traced from the right on pages white' was a conscious oddity of priests. Port was circulated deasil; to send it round widdershins was extremely unlucky. Playing cards are dealt deasil; we turn screws deasil. It is possible to trace our right-handedness to the perception of deasil as the normal direction, and widdershins as abnormal, topsy-turvy, unlucky. The one striking instance I can think of where widdershins is regarded as the normal direction is in the mathematical measurement of angles where the measurement is made counter-clockwise. The reason for this exception probably is that the measurement originates at the centre of the circle; there is the normal measurement from left to right for the positive direction, and then, almost instinctively, we measure upward for the positive direction, and so get the counter-clockwise rotation.

When the compass is divided into degrees, measurement starts at the top, there is no disturbing circumstance, and the direction of measurement is deasil.

The fact that civilisation originated in the northern hemisphere explains why we regard deasil as normal. If it had originated in the southern hemisphere, we might have had everything reversed. We should see the sun move from east to west through the north, that is from right to left. The direction we call widdershins would be normal; deasil would be odd, outré, ill-omened, contrary to the course of the sun and the normal widdershins clock. The normal screw would be left-handed. The writing of priests would be 'traced from the left by fingers deft'. Most of us would be left-handed; our mentors would say: 'No, dear, not that hand; the knife in the left hand.'

The deasil movement of the sun across the sky every day is obvious; the most obtuse could hardly avoid seeing it. A very slight study of the moon and the stars is sufficient to show that they also appear to move deasil round the sky. The circumpolar stars, including the stars of the Plough, may be seen to describe complete circles, deasil, round the Pole Star. The apparent deasil rotation is obvious enough. It took a considerable effort of reasoning to perceive that the apparent movement is the effect of a real rotation of the earth in the opposite direction. The heavenly bodies appear to move deasil because the earth is rotating widdershins.

There can be no reasonable doubt about the rotation of the earth; the alternative theory of the rotation of the sun, moon and stars about the earth involves too many improbabilities. We should have to accept the extraordinary idea that each star describes a circle round the earth every 24 hours (or a little less). The distance of Arcturus is about 800 billion miles; the length of its circular path would be $2\pi$ times that distance, or about 5000 billion miles. Arcturus would have to travel that distance in 24 hours, at a speed of nearly 60,000 million miles a second; which is at least highly improbable, and to most people far beyond the bounds of credibility. There are stars from which light takes thousands and even millions of years to come; we should have to accept the fact that these stars perform six times the journey in 24 hours. On top of this we should have to accept the fact that these improbable speeds are all in exact proportion to the star's distance from the earth. We should be driven to accept one of two extraordinary ideas; either that the earth happens to be situated at the exact centre of a universe revolving in circles about it at incredible speeds; or else that the whole universe is governed by the earth, which has a mere 300,000th part of the mass of the solar system, which is itself an insignificant part of the universe.

There is nothing for it but a widdershins earth.

The idea of a stationary earth with the sun revolving round it once every year is also supported by appearances. We can see by observation of the stars that the sun performs a yearly journey through the signs of the zodiac. This apparent movement is from right to left, widdershins.

The diagram shows why this should be so. $S$ is the sun, with the earth's nearly circular orbit drawn round it. When the earth is at $A$ the sun is seen in the direction $AC$; when the earth is at $B$ it is seen in the direction $BD$. That is, while the earth moves from $A$ to $B$ the sun moves from $C$ to $D$. The sun's apparent movement round the zodiac is in the same direction as the real revolution of the earth round the sun, widdershins.

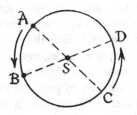

There is one real movement in the skies that can be readily observed: the widdershins revolution of the moon round the earth. There is actually a double movement of the moon: the apparent, deasil, movement which causes rising and setting; and the real, widdershins, revolution round the earth which causes the moon to perform a monthly journey round the zodiac. At new moon the moon may be seen a little to the left of the setting sun. It moves farther and farther to the left of the sun, until at full moon it is 180° away from the sun, that is at the opposite side of the sky. It continues to move to the left, and completes the ellipse round the earth in about $27\frac{1}{3}$ days. The daily movement of the moon is:

$$\frac{360°}{27\frac{1}{3}} = \text{about } 13° \text{ or rather more.}$$

A picture of 5° in the sky may be obtained by looking at the Pointers of the Plough, which are about 5 degrees apart. The daily widdershins movement of the moon is a little over $2\frac{1}{2}$ times the distance between the Pointers. The moon itself has an angular width of about half a degree; so the moon moves its own width in:

$$\frac{24 \text{ hours}}{26} = \text{about } 55\frac{1}{3} \text{ minutes.}$$

Astronomers cannot help regarding the solar system as a small widdershins universe, because the real movements with which they are familiar are nearly all widdershins: the rotation of the sun on its axis, the rotation and revolution of the earth and the moon, the rotation and revolution of the planets and of most of their satellites. It is only amongst the satellites of the outer planets that deasil movements are found. The idea that widdershins movements are normal is so thoroughly ingrained in astronomers, contrary to popular feeling on the subject, that deasil revolutions are usually called 'retrograde' and regarded as anomalies to be explained.

The almost invariable recurrence of the same kind of rotation and revolution throughout the solar system, and the fact that variations from this normal occur only in the outer parts of the system, have been taken as a confirmation of the idea that the whole system had an original widdershins spin. According to this view the solar system has been widdershins from its beginning.

Great fleas have little fleas upon their backs to bite 'em,
Little fleas have lesser fleas and so ad infinitum.

This attractive idea has been applied in the reverse direction to the universe at large. The moon goes round the earth; the earth goes round the sun; the sun goes round—what? It has been suggested that the sun, together with other stars, is revolving in the same way about some remote centre, perhaps the centre of gravity of the whole galactic universe.

We are seldom told about the straightness of the circumference of such an orbit. In the small diagrams we use, the curvature is obvious, and it looks as if we should be able to measure it with ease. But the sort of centre that has been suggested may be 10,000 parsecs away, or more, and that is in the region of 200 thousand billion miles.

Here is a drawing of such an orbit, and there is no doubt about the curvature; it is plain to see. $P$ is a point on the orbit, and the sun is at $P$. $O$ is the remote centre, 200 thousand billion miles away, round which the sun is circling. $PA$, the tangent at $P$, is the straight path on from $P$; and $PB$ is the curved path along the orbit. We want to know the length $AB$; that is, how far the revolving sun has to fall toward the distant centre in order to keep on the curved orbit.

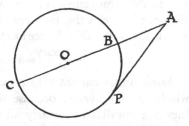

We know from geometry that:

$$AP^2 = AB \cdot AC.$$

Suppose we want to find the fall when $AP$ is a million miles. We know that $AB$ is very small compared with $BC$; and so we can say that $AC = BC = 400{,}000$ billion miles. And now we have:

$$1{,}000{,}000^2 = AB \times 400{,}000 \times 10^{12}$$

or

$$10^{12} = 400{,}000 \times 10^{12} AB$$

and

$$AB = \frac{1}{400{,}000} \text{ mile}$$
$$= \text{about } \tfrac{1}{6} \text{ inch.}$$

So that the deviation from a straight line is about a sixth of an inch in a million miles. That is to say, a drop of a sixth of an inch in a million miles would be sufficient to change a straight course into a circular or elliptical orbit round the remote centre.

Compared with the nearer stars, it has been computed that the solar system is travelling more or less toward distant Vega. The calculated speed is 250 kilometres per second, or nearly 5000 million miles in a year. Let us see what the deviation from a straight course must be in a year. The equation is the same as before, except that we now have on the left the square of 5000 millions. The equation is:

$$5000^2 \times 10^{12} = 400{,}000 + 10^{12} AB,$$
$$AB = \frac{5000^2}{400{,}000}.$$
$$= 62\tfrac{1}{2} \text{ miles.}$$

The deviation from a straight path in the course of a year is no more than 62 miles or so in about 5000 million miles.

Let us go a stage further and think of a whole century of observation. The movement of the solar system in that period is 500,000 million miles. The equation becomes:

$$500,000^2 \times 10^{12} = 400,000 \times 10^{12} AB,$$
$$AB = \frac{500,000^2}{400,000}$$
$$= 625,000 \text{ miles.}$$

That is to say, in a century of observation we should have a drop of five-eighths of a million miles from a straight course of half a billion miles. This is one part in 800,000, or $\frac{1}{8000}$ of 1 per cent.

The length of the complete orbit is:

$$2\pi r = 2\pi \times 200,000 \times 10^{12}$$
$$= 1 \cdot 26 \times 10^{18} \text{ miles,}$$

or $1\frac{1}{4}$ trillion miles.

The time taken to make a complete revolution in the orbit would be:
$$\frac{1 \cdot 26 \times 10^{18}}{5000 \times 10^6} = 2 \cdot 52 \times 10^8 \text{ years,}$$

or 252 million years.

This is not to say that the solar system is not revolving about some remote centre, but only to point out one of the difficulties in obtaining direct evidence of such a revolution and of deciding where the remote centre is. After all, exactly the same problem, though on a much smaller scale, arose in showing that the earth is an approximate sphere. So long as observation was limited and measurements short there was little to go on. We have seen that a drop of two inches in the mile is sufficient to change the flat earth into a sphere. It was not until measurements were extended over several degrees of latitude that there was any certainty. In the rotation of the solar system, for a single second of the arc of rotation we should have to wait:

$$\frac{252 \text{ million years}}{360 \times 60 \times 60} = \text{about 200 years.}$$

The yearly angular movement is just about the limit of what is measurable.

It appears that the axis of revolution of the solar system is, more or less, perpendicular to the ecliptic, and so to the whole solar system. The revolution is deasil, so it appears that the solar system is a small widdershins swirl in a larger deasil universe.

# CHAPTER 5

## The All-Composing Hour

THE HOUR is the most firmly entrenched of units. It is the universal unit of time, the one unit that even ten-ridden decimalists do not propose to change. It is too convenient a unit for times of work and play and sleep to be readily given up. It has had a strange history as a unit. Until the eighteenth century an hour was the twelfth part of the time between sunrise and sunset, or between sunset and sunrise. 'The long hours of a winter's night' literally were long, something like eighty minutes long; just as the 'long summer hours' were something like eighty minutes of daylight.

It was a picturesque way of measuring time. It died out when the gradual spread of clocks made the exact measurement of time possible to all. Only the astrologers, the necromancers of a dead cult, retain the unequal hour. In the old days many astronomers turned astrologer in their play hours, and raised money by casting horoscopes for the credulous; at such times they used the unequal hour. But when they turned to serious observation of the stars they made their calculations in 'equinoctial hours', which are the hours we use now.

It is commonly thought that the earth rotates on its axis 365¼ times in a year, and that an hour is the 24th part of a single rotation. This is not so, and the diagram shows why. The point $A$, on the earth, is in the noon position, directly under the sun. An exact number of rotations would carry the point $A$ to the position $B$, where the earth has moved on in its orbit. At $B$

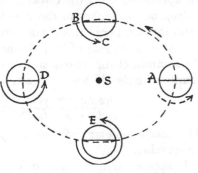

the point is not in the noon position; it has to move the additional distance $BC$; this is a quarter of a rotation, so that noon is actually six hours later. At $D$ the additional rotation has accumulated up to half a rotation. At $E$ it has reached three-quarters of a rotation; and back at $A$ the additional rotation has reached a complete rotation.

Thus it arises that in a year, which is the time of a complete revolution of the earth round the sun, there are $366\frac{1}{4}$ rotations, and not $365\frac{1}{4}$. And thus also the solar day, which is the time from noon to noon, is a little longer than the time of a rotation.

There is no such complication when we measure the time of rotation by reference to the distant stars. The rotation of the earth brings the stars in turn across the meridian; and after an exact rotation a star again crosses the meridian. This period is called a *sidereal day*. It is the exact time of the rotation of the earth as measured by the apparent rotation of the stars.

$365\frac{1}{4}$ solar days is a fairly exact measure of the solar year, so we can find the length of the sidereal day by dividing this time by $366\frac{1}{4}$.

$$\frac{365\frac{1}{4} \text{ mean solar days}}{366\frac{1}{4}} = \frac{365\frac{1}{4} \times 24}{366\frac{1}{4}} \text{ hours}$$
$$= 23 \cdot 93447 \text{ hours}$$
$$= 23 \text{ hours } 56 \text{ minutes } 4 \text{ seconds.}$$

That is the length of a sidereal day in solar hours, minutes and seconds, to a pretty close degree of approximation.

Astronomers use the sidereal day, with its divisions into sidereal hours, minutes and seconds, because of its convenience for their work. To fix the position of a star we need two measurements similar to latitude and longitude for earth measurements. If we imagine the plane of the earth's equator extended outwards indefinitely, we have the *celestial equator*.

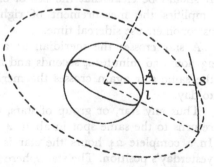

Measurements are made north and south from this circle, just as measurements of latitude are made on the earth. The angle $l$, for instance, is the latitude of a place $A$ on the earth's surface, or the celestial latitude of a star in the direction $AS$. Celestial latitudes, measured north or south from the celestial equator as zero line, are called *declinations*.

To measure star longitudes we need an arbitrarily chosen zero, like the meridian of Greenwich for earthly longitudes. The meridian chosen is the one which goes through the *first point of Aries*; this is, or was, the most westerly point of the constellation of Aries, the Ram. It is called the first point, because the sun, in its apparent revolution of the heavens from right to left, enters the constellation from the west. (There is no trouble about the directions east and west because the constellations of the zodiac are not visible when in the north.) The first point of Aries crosses the meridian at zero hours. Celestial longitude, measured eastward from the first point of Aries and its meridian, is called *right ascension*.

The name 'first point of Aries' is still retained for the vernal equinox, one of the two points where the earth's orbit cuts the plane of the equator, in spite of the fact that precession is carrying the point westward at the rate of about 50″ yearly. Thus the right ascensions of stars increase by about 50″ yearly.

At zero hours the first point of Aries crosses the meridian. 3 hours 18 minutes 9 seconds later by sidereal time a certain star is observed to cross the meridian. Its right ascension is 3° 18′ 9″; or, as more commonly expressed, it is 3 hours 18 minutes 9 seconds. It should be clear that the use of clocks showing sidereal time simplifies the measurement of right ascension. That is why astronomers use sidereal time.

A star crosses the meridian at a certain time. Punctually 23 hours 56 minutes 4 seconds and a fraction of a second later, the same star again crosses the meridian; and so on from day to day.

Thus any star, or group of stars, completes a revolution and returns to the same spot in about 4 minutes less than 24 hours. In a complete 24 hours the star is about 4 minutes ahead of yesterday's position. The star sphere therefore appears to revolve from left to right, deasil, at the rate of about 4 minutes in 24 hours. This is another way of describing the apparent widdershins revolution of the sun through the signs of the zodiac.

In 4 minutes less than 24 hours the stars appear to revolve through 360°. This is:

$$15° \text{ in } 1 \text{ hour,}$$
and $$1° \text{ in } 4 \text{ minutes.}$$

Hence the star sphere appears to revolve from left to right a little less than 1° every 24 hours. It completes the revolution in a year.

It is a curious point that the Chinese divided the circle into 365¼ degrees. Thus the average daily movement of the earth in its orbit, or of the sun through the signs of the zodiac, or of the star sphere, is one Chinese degree. It must have been a very awkward division for Chinese instrument makers to make, but the Chinese never seem to have boggled at that kind of difficulty. Our familiar 360 degrees probably came from Babylon; it was introduced into Greece by Hipparchus. Quite possibly 360 was chosen as the nearest whole number to 365¼ that has good divisibility. If so, it was a very fortunate choice, especially in divisibility by 3 that gives the angles of 60° in the equilateral triangle.

We speak familiarly of G.M.T. or Greenwich Mean Time. But why 'mean time', why not just 'Greenwich time'. Anyone who has carefully observed a sundial from day to day will know the answer. A properly set sundial shows exact solar time, and in particular it gives the exact solar noon. Sundial time differs by varying amounts from Greenwich Mean Time as pipped out by the B.B.C.

The reason lies eventually in the ellipticity of the earth's orbit. We know that the earth moves round the sun in an ellipse, with the centre of the sun at one focus of the ellipse. Hence the distance of the earth from the sun varies from day to day. The pull of the earth and the sun on each other is greatest when they are nearest, and the earth then moves at its greatest speed in its orbit. The exact law was given by Kepler: the line joining sun and earth sweeps over equal areas of the orbit in equal times.

The diagram shows the orbit of the earth, with the ellipticity greatly exaggerated. Between $A$ and $B$ the earth is at its nearest to the sun; at $C$ and $D$ the earth is at its farthest. In moving from $A$ to $B$ the line $AS$ sweeps over the area $ASBEA$; in moving from $C$ to $D$ the radius sweeps over the area $CSDFC$. If the two areas are equal, then the earth takes as long to move from $C$ to $D$ as it does to travel the greater distance from $A$ to $B$.

If we make the two areas very narrow, then we can treat them as if they were narrow triangles. The heights of these narrow triangles are $SE$ and $SF$. The equal areas are:

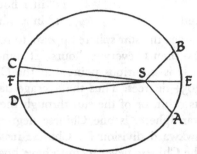

$$\tfrac{1}{2}AB.SE = \tfrac{1}{2}CD.SF.$$

Let us take $SE$ as the shortest (perihelion) distance, and $SF$ as the greatest (aphelion) distance. $SE = 91 \cdot 34$ million miles and $SF = 94 \cdot 46$ million miles.

$$AB.SE = CD.SF.$$

So
$$\frac{AB}{CD} = \frac{SF}{SE}$$
$$= \frac{94 \cdot 46}{91 \cdot 34}$$
$$= 1 \cdot 034.$$

That is a difference of 34 parts in a thousand or $3 \cdot 4$ per cent. The greatest speed of the earth in its orbit is nearly $3\tfrac{1}{2}$ per cent greater than the least speed.

We have seen that the length of the solar day, that is, the time from solar noon to solar noon, depends on the speed of the earth in its orbit. The farther round the earth moves between one noon and the next, the greater is the time between noon and noon. So that when the earth is nearest the sun, and therefore moving quickest, days are apt to be longer than when it is farther away.

There is another thing that affects the length of the solar day. The sun's apparent yearly path through the skies (the ecliptic, or plane of the earth's orbit) is inclined to the plane of the equator. Every day the sun travels about one degree along the ecliptic. At the equinoxes a distance $AB$ on

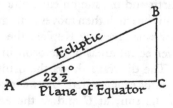

the ecliptic is longer than $AC$ on the equator. There must be a greater angular turn along the ecliptic in order to have a turn of one degree (corresponding to a day) along the equator. At the equinoxes the length of a solar day is greater than the mean day. The difference is made up at the solstices, when the solar day is correspondingly shorter.

Thus we have the varying speed of the earth in its orbit tending to produce longer days at perihelion (January) and shorter days at aphelion; and the inclination of the ecliptic tending to produce longer days at the equinoxes and shorter days at the solstices. The combination of these two causes gives us solar days of varying length throughout the year.

The mean solar day is the average length of the solar day throughout the year, that is, the length of a year divided by $365\frac{1}{4}$. The purpose of this average is to enable clocks to maintain a regular rate of movement without the necessity of having to be corrected daily so as to coincide with sundial time.

The difference between solar time and mean time is called the *equation of time*. Almanacks give the number of minutes and seconds that have to be added to or subtracted from sundial time (or time obtained by observation of the sun) in order to obtain mean time as shown by an accurate clock. Clocks show the same time as sundials on four days of the year, on or about 16 April, 15 June, 1 September and 25 December. The greatest difference is on or about 4 November, when about $16\frac{1}{2}$ minutes has to be subtracted from sundial time to find mean, or clock, time; sundial time is then $16\frac{1}{2}$ minutes fast on clock time. The other three maxima, between the dates when the two times coincide, occur on: 12 February, when sundial time is about $14\frac{1}{8}$ minutes slow; 15 May, when sundial time is $3\frac{3}{4}$ minutes fast; 27 July, when sundial time is $6\frac{1}{8}$ minutes slow.

During the month of January sundial noon comes later and later day by day (as compared with clock noon). Sunrise in London moves back from 8 minutes past 8 to 17 minutes to 8 during the month, a difference of 25 minutes. Sunset moves on from 1 minute to 4, to 16 minutes to 5, a difference of 45 minutes. The difference of 20 minutes between the two changes is due to an advance of 10 minutes in solar noon compared with clock noon. It is a little bit of uncovenanted 'daylight saving'.

In September there is almost exactly the same difference, but in the opposite direction. Between the first and last days of the month the sun rises 46 minutes later and sets 66 minutes earlier. The difference between the two changes is due to a backward movement of 10 minutes in sundial noon compared with clock noon. September is the month when 'days begin to close in'.

We have to say 'on or about' when referring to the dates because the odd quarter of a day in the year introduces variations of 6 hours, 12 hours, 18 hours, and then back to zero. In the first year after leap year we start the year 6 hours too soon, in the second year 12 hours too soon, and in the third year 18 hours too soon. So that calendar dates for such annual events as the longest and shortest days are apt to vary from year to year.

There was a pseudo-religious objection to the institution of 'summer time' by people who objected that it was tampering with 'God's time'. What the objectors apparently meant was Greenwich Mean Time, and that, as we have seen, is a highly artificial method of reckoning.

As an added element of artificiality we have Standard Time. The purpose of Standard Time is to avoid, wherever possible, the inconvenience of having two or more times in the same country, and also to make calculations as simple as possible by having the time zones differ by exact hours, or in a few cases by half hours.

When we divide the 24 hours round the 360 degrees of the compass of the whole world, we have:

$$360° = 24 \text{ hours,}$$
$$1 \text{ hour} = 15°,$$
$$1° = \tfrac{1}{15} \text{ hour} = 4 \text{ minutes.}$$

For purposes of Standard Time the world is divided into 24 zones each 15° wide, with adjustments to include the whole of one country in one time zone wherever this is possible. Down the middles of the time zones run the meridians 0°, 15° east and west, 30° east and west, and so on. These meridians differ by periods of one hour, and give the times to the zones.

The British Isles, France, Belgium, Spain, and Portugal, all have Greenwich time. This time zone extends from about 8° E. to 9° W., so that there is an extreme difference between the eastern and western limits of 17 degrees, or $\frac{17}{15}$ hours = 1 hour 8 minutes. If they keep the same clock hours, dwellers in the west of this time zone get up 1 hour 8 minutes earlier by sun time than dwellers in the east of the zone. Even in the restricted area of Great Britain there is a difference of about $5\frac{1}{2}°$ of longitude between Greenwich and Land's End. $5\frac{1}{2}°$ is equivalent to 22 minutes, so that the hands of clocks in Cornwall are permanently set forward about 20 minutes fast on local time. So that in Cornwall, as compared with London, there is 20 minutes 'daylight saving' all the year round. This is very noticeable in early autumn, when to Londoners who visit Cornwall it seems as though 'night will never come'.

Europe has three time zones: Greenwich, Mid-European, and East European. Mid-European time is one hour fast on Greenwich; it is the same as Greenwich time when we have 'summer time' (one hour fast), and Mid-Europe has normal time. Mid-European time is the standard time from Scandinavia southward to Italy. Holland stands outside the Standard Time Zone agreement. Its official time is 19 minutes 32·1 seconds fast on Greenwich. That represents a longitude of $19'\ 32 \cdot 1'' \times 15 = 4°\ 53'$ East which is just about the longitude of Amsterdam. East European time is two hours fast on Greenwich; it is the official time of Finland, European Russia, Greece, Turkey, Rumania, Bulgaria, and Estonia.

If there is any truth in the proverb 'Early to bed, and early to rise, makes a man healthy, wealthy, and wise', it should certainly apply to sundial time and not to the highly artificial Standard Time. There should be a distinct, or at least an observable, difference in health, wealth and sagacity between the eastern and western limits of the time zones. Cornwall should be a little healthier, wealthier and wiser than London, the west of Ireland more so, and Portugal almost as wise as Ireland. Norway has a considerable extension in degrees of longitude; the extremes are 25° apart, and that makes a time difference of $1\frac{2}{3}$ hours. So the west of Norway should be $1\frac{2}{3}$ hours healthier, wealthier and wiser than the east. North America has five time zones: Atlantic, Eastern, Central, Mountain and Pacific. There is ample room across the wide longitudes of North America for an inquiry into the relative health, wealth and wisdom of people living in the extreme east and the extreme west of the time zones. School, office, and factory hours would show whether the people of the west really were getting up early, or whether they were fudging by making the opening hours later; the practice on farms would need to be carefully watched. The health authorities would no doubt be ready to supply the necessary data for comparing health in different parts of the zones; the banks would be able to assess wealth; and I have no doubt that psychologists would be ready to supply Sagacity Tests to estimate the most intangible and unpredictable of human faculties.

A fearful thought has just occurred to me. Perhaps it is people who get up earlier than their neighbours who count. I must say that so far all I have observed in them is an insufferable arrogance.

People sometimes confuse the shortest day with perihelion, the point of closest approach between earth and sun. Indeed I have an atlas which puts them together on 21 December. Actually they have nothing to do with each other; the shortest day, for northern latitudes, occurs when the North Pole is most completely turned away from the sun; perihelion is simply a point on the earth's orbit.

It just happens that the two are in the same part of the year. The shortest day is 21 December. Perihelion is reached about 3 January. This is the part of the orbit where the earth is moving quickest—so as to shorten northern winter. The earth is at aphelion, the most distant part of its orbit, about 5 July. It moves more slowly over this part of its orbit—so as to lengthen northern summer.

In this diagram the ellipticity of the earth's orbit has been exaggerated, in order to draw attention to the fact that the part BCA, which includes perihelion and northern winter, is shorter than ADB, which includes aphelion and northern summer.

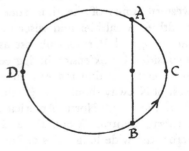

There are two reasons then why the northern winter part of the year should be shorter than the summer part. From the autumnal equinox (23 September) to the spring equinox (22 March) is only about $178\frac{1}{2}$ days; whereas from the spring equinox to the autumnal equinox, the period of northern summer, the interval is $186\frac{1}{2}$ days.

There is something else we have to include when striking a balance between north and south. The inverse square law shows that when at perihelion the earth receives:

$$\frac{94 \cdot 46^2 \text{ (square of aphelion distance)}}{91 \cdot 34^2 \text{ (square of perihelion distance)}}$$

$$= 1 \cdot 069 \text{ times as much heat as when it is at aphelion.}$$

That is, nearly 7 per cent more.

The Northern Hemisphere has a longer summer in the cooler part of the orbit, and a shorter winter on the warmer part of the orbit. Both conditions tend toward a milder climate for the Northern Hemisphere. The Southern Hemisphere has a shorter summer in the warmer part of the orbit, and a longer winter in the cooler part of the orbit. The tendency is toward a more extreme climate—hotter summers and colder winters. The distribution of land and water introduces complications; we notice especially that land in the Southern Hemisphere is almost entirely confined to tropical and subtropical regions. The extreme nature of southern winter is felt most strongly in the Southern Continent; in the long cold winter there is a great extension seaward of the icecap that rings that inhospitable continent.

We have seen that the sidereal day is the exact time of rotation of the earth as measured by reference to the stars. Corresponding to the sidereal day there is the *sidereal year*. This is the exact time of a revolution of the earth about the sun, as measured by reference to the stars. At the end of a sidereal year every star has returned to the exact spot that it occupied in the sky at the beginning of the year. The length of a sidereal year is $365 \cdot 25636$ mean solar days, or $366 \cdot 25636$ sidereal days—just a little more than $365\frac{1}{4}$ mean solar days.

There is another kind of year which is called a *solar* or *tropical* year. The thing we are interested in throughout the year is the seasons.

We want the seasons to come in the same place in each year. In particular we want the spring equinox to come in the same place in the calendar each year, and that is what it will not do, unless we are careful with the calendar.

The trouble is *precession*.

The gyroscope has so many uses in these days that we have become familiar with the idea of precession. If we press on the axis of a gyroscope it begins to precess; that is, the ends of the axis begin to move round in circles. The rotating earth is a kind of gyroscope; the pull of the sun and the moon on the equatorial bulge exerts a pressure that causes the earth to precess. The North and South Poles move slowly round in circles.

The diagram shows the earth, looked down at from the north. The North Pole is shown 23½° out from the perpendicular to the plane of the earth's orbit. The pole precesses very slowly, that is, it moves round in a circle. The precession is deasil, or as astronomers say 'retrograde', that is, opposite to the general movement of the solar system.

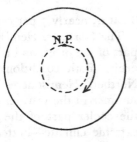

The spring equinox occurs when the earth's axis is not inclined toward the sun. At *A* we have the spring equinox when the earth is near perihelion. When the end of the axis has done a quarter turn, we have the spring equinox at *B*. A half turn takes it to aphelion at *C*; a three-quarters turn takes it to *D*; and a complete turn back again to perihelion at *A*.

Thus from spring equinox to spring equinox is a little less than a sidereal year. The equinox *precedes* the completion of the orbit; hence the name precession.

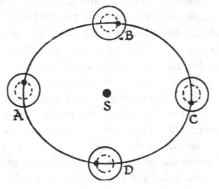

The rate of precession is very slow. The equinox retrogresses about 50·26″ each year; this angle is called the *constant of precession*, though actually precession is rather variable. To complete the circle would take:

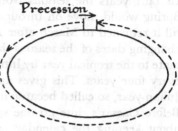

$$\frac{360°}{50·26″} = \frac{360 \times 60 \times 60}{50·26} \text{ years}$$

$$= 25,780 \text{ years}.$$

This period, in which the equinoxes make a complete circuit of the orbit, is called a *Platonic* or *great year*. The solstices, winter and summer, move round the orbit at the same rate. We have seen that the winter solstice now comes about 12 days before perihelion. This gap will slowly increase, until after about 12,000 years the winter solstice will occur at aphelion.

In a tropical year the earth travels a little less than a complete revolution. The angle is:

$$360° - 50·26″ = 359° \; 59' \; 9·74''$$
$$= 1295949·74''.$$

So that the length of the tropical year is:

$$\frac{1295949·74''}{360°} \text{ of the sidereal year}$$

$$= ·999961 \text{ of the sidereal year}.$$

The length of the sidereal year is 365·25636 mean solar days. So the length of the tropical year is:

$$365·25636 \times ·999961 \text{ mean solar days}$$
$$= 365·24212 \text{ mean solar days}.$$

The difference between the two years is ·01424 day.

If we were to use the sidereal year for our calendars, the spring equinox would come ·01424 day earlier each year, and that is about $\frac{1}{70}$ of a day. In 70 years the spring equinox would come a day earlier in the calendar. In ten centuries it would have worked back 14 days.

That is why calendar makers have to do the best they can with the tropical year of 365·24212 days.

The *civil year* or *calendar year* consists of course of an integral number of days. A year of 365 days would soon put the seasons out of place. The spring equinox would come about 6 hours later each year, a day in four years, and about 24 days in a century. In 1461 years the seasons would almost have circled the year. Spring would move on through March, April, May, and so on, till it returned to March after 1461 years. In order to avoid the changing dates of the seasons the civil year was made to approximate to the tropical year by introducing the extra day of leap year every four years. This gives an average year of $365\frac{1}{4}$ days—the Julian year, so called because it was introduced by Julius Caesar. Before Caesar's time no one seems to have worried very much about keeping the calendar exact. The ancient Egyptians, for example, had a year of 365 days, so that their seasons actually did circle the year in 1461 of their short years.

Caesar's calendar was excellent. In it the months had alternately 31 and 30 days; February (the last month of the year) had 29, except in leap year. Augustus Caesar made a mess of Julius Caesar's work. He stole a day from February to add to his own month of August; and then, because three thirty-one day months came together, he had the extra days moved from September and November and added to October and December.

The Julian year is a little too long. The difference is:

$$365 \cdot 25 - 365 \cdot 24212 = \cdot 00788 \text{ day.}$$

In a hundred years this amounts to $\cdot 788$ day, or just over $\frac{3}{4}$ of a day. So in every 400 years three leap years are omitted (the omitted leap years are the exact hundreds, except where these are multiples of 400). This gives an average year of $365 \cdot 2425$ days, which is still a little too long. Perhaps in the rather distant future another leap year will be omitted—perhaps the year 4000.

Reformers have dealt drastically with years in order to bring the spring equinox back to its place in March. Julius Caesar ordained 445 days for the year before his calendar began. When Pope Gregory reformed the calendar in 1582 he ordered ten days to be omitted so as to bring the equinox back to 21 March. When the New Style was introduced into England eleven days had to be omitted from the calendar; an extra day, because 1700 had been counted as leap year in the English calendar, but not in the .

Gregorian. The day following 2 September 1752 was counted as 14 September, and a puzzled populace demanded with menaces 'give us back our eleven days'.

It is a curious fact that the theatrical profession celebrate Shakespeare's birthday eleven days after 23 April, that is, on 4 May. If they really want to preserve the Old Style date, it should now be 6 May, since two years have transpired since the change, 1800 and 1900, that would have been leap years in the Julian calendar, but not leap years in the Gregorian calendar. However, they are at least keeping Shakespeare's birthday at the point in the cycle of seasons where it properly belongs.

The month is an even odder thing than the year. The most obvious kind of month is the period from new moon to new moon—the *synodical month*. The average length of this month is about 29½ days, so that twelve of them come not far short of a year: 354 days out of 365¼. This probably accounts for the all but universal division of the year into twelve *calendar months*, usually of 30 or 31 days. This division extended literally 'from China to Peru', westward. Only in Mexico does there seem to have been a departure from the general practice. The year was divided into 18 months of 20 days, with 5 added days to make up the 365. Allowance was made for leap year by adding 13 days at the end of each cycle of 52 years. Mexico also seems to stand alone in having had a five-day week, instead of the all but universal seven.

There is a *solar month* which is the twelfth part of a solar year. But the oddest of all months is the *lunar month*, so called because it has no sort of exact relation to any moon period. The lunar month has 28 days exactly, so that its only close relative is the seven-day week. The lunar month has had considerable vogue both in law and in textbooks on arithmetic. The table ran smoothly from '60 seconds = 1 minute' to '4 weeks = 1 lunar month', and then it collapsed at the fantastic year.

As we should expect from the rapid rate of revolution of the moon, there is a considerable difference between the synodical month (new moon to new moon) and the *sidereal month*, which is the time of an exact revolution of the moon round the earth.

We can get a fair idea of the difference by observing that in $29\frac{1}{2}$ days the earth has moved through:

$$\frac{29\frac{1}{2}}{365\frac{1}{4}} = \frac{118}{1461} \text{ of its orbit.}$$

$$360° \times \frac{118}{1461} = \text{about } 29 \cdot 07°.$$

So that between new moon and new moon the earth has moved forward about $29 \cdot 07°$ in its orbit. The diagram shows that the moon has to revolve $360°$ plus this angle. A single revolution would take it to $A$; the additional $AB (= 29 \cdot 07°)$ brings it again into the new moon position at $B$.

$$360° + 29 \cdot 07° = 389 \cdot 07°.$$

29 days $12\frac{3}{4}$ hours is a more exact measure of the synodical month. The moon revolves:

$$389 \cdot 07° \text{ in 29 days } 12\frac{3}{4} \text{ hours,}$$

and $\qquad 360° \text{ in } \dfrac{360}{389 \cdot 07} \times 29 \text{ days } 12\frac{3}{4} \text{ hours}$

$$= \frac{360}{389 \cdot 07} \times 708\frac{3}{4} \text{ hours}$$

$$= 655 \cdot 8 \text{ hours}$$

$$= 27 \text{ days } 7 \cdot 8 \text{ hours.}$$

That is a very fair result considering the nature of the approximations. The actual length of the sidereal month (the exact time of a revolution of the moon) is a very little less. It is very nearly 27 days $7 \cdot 72$ hours, or 27 days, 7 hours 43 minutes. Unless very exact calculations are being made the length of the sidereal month may be taken as $27\frac{1}{3}$ days.

# CHAPTER 6

## *Ellipses and Orbits*

ELLIPSES PLAY an important part in the geography of the solar system. We have seen that the meridians of the earth are ellipses; it is parts of ellipses that have to be measured, and fitted on to the oblate spheroid shape, when the size of the earth is being measured. The ellipticity is small, it is only $\frac{1}{297}$ or about $\frac{1}{300}$; but it has to be allowed for. The ellipticity of Mars is thought to be $\frac{1}{270}$, though the measurement is difficult and the result doubtful. The measurement is good enough however to show that the two planets are very much alike in shape, and that both are almost exact spheres.

Of all the planets Saturn is the most elliptical; it has an ellipticity of $\frac{1}{9}$, which is about 33 times as great as the ellipticity of the earth. The difference between equatorial and polar radii of Saturn is $\frac{1}{9}$ of the equatorial radius. Saturn is an oblate spheroid, with an ellipticity that is visible to the eye. Jupiter is another planet with a distinct ellipticity; it is $\frac{1}{17}$, or about half the ellipticity of Saturn.

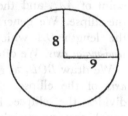

The ellipticity of a planet is of importance chiefly to possible inhabitants; and it is a confirmation of the oblate spheroid idea imagined for the earth, to find it shown by other planets. But the chief importance of ellipses in the solar system lies in the fact that the orbits of the planets and their satellites are all ellipses. If we want to understand these orbits, we have to know something about ellipses; and it happens, very fortunately, that the essential facts are easily arrived at.

Most people know the method of constructing an ellipse with two pins, a loop of thread, and a pencil. We spread out a sheet of paper; then we fix the pins upright, put the loop of thread over them, draw it out taut with the point of a pencil, and draw round.

We can deduce what it is necessary to know about an ellipse from this method of drawing it.

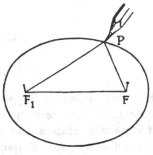

The two points where the pins are fixed are called the *foci* of the ellipse. $P$ is any point on the ellipse we are drawing. The length of the loop is $FF_1 + PF + PF_1$, and this does not vary; $FF_1$ also does not vary. So $PF + PF_1$ is the same for every point on the ellipse. The sum of the distances from any point on the ellipse to the two foci is constant.

We continue the line $FF_1$, joining the two foci, right across the

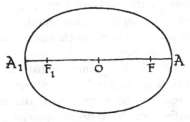

ellipse. This gives us the line $AA_1$, which is called the *major axis* of the ellipse. It divides the ellipse into two equal and symmetrical parts. $O$ is the midpoint of $AA_1$, and the *centre* of the ellipse. We generally want the length $OA$ which is the semi-major axis. We often write $a$ for the length $OA$.

We draw $BOB_1$ at right angles to $AA_1$. $BB_1$ is called the *minor*

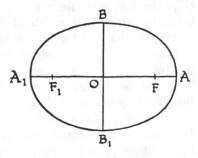

axis of the ellipse. $BB_1$ also divides the ellipse into two equal and symmetrical parts. $OB$ is half $BB_1$, and so it is the *semi-minor axis*. We often write $b$ for the length $OB$. The lengths $a$ and $b$ are very important in describing an ellipse.

It is instructive to draw ellipses with the same loop, and with the foci at different distances apart. When the foci are brought together, the ellipse becomes a circle of radius $a$. The more the foci are spread out the narrower it becomes, until finally it becomes a straight line.

The drawing should show that the shape of the ellipse depends on what fraction $OF$ is of $OA$. $OF$ cannot of course be greater than $OA$; otherwise we could not put the loop over the pins. It is always a fraction of $OA$, and the greater the fraction the more elongated the ellipse is. This fraction $\dfrac{OF}{OA}$ is called the *eccentricity* of the ellipse; it measures how much $F$ (and $F_1$) are out from the centre $O$. If for example the eccentricity is $\frac{1}{3}$, then $F$ is $\frac{1}{3}$ of $OA$ out from $O$. We usually write $e$ for the eccentricity. $e$ is always a fraction less than one.

Let us see what we can do with these few facts.

The semi-major axis ($a$) and the eccentricity ($e$) describe an ellipse completely; if we are told these quantities, we can draw the ellipse. Suppose we are given: $a = 2$ inches, $e = \frac{1}{3}$, and we want to draw the ellipse. We draw $AA_1 = 4$ inches, and bisect it at $O$. We make $OF = \frac{1}{3}OA$, and $OF_1 = OF$. We now have the two foci, $F$ and $F_1$. We fix pins at $F$ and $F_1$. Then we make a loop of thread to stretch from $F_1$ to $A$. We put this loop over the pins, draw it out taut, and use it to draw the ellipse. This ellipse fulfils the conditions of having $a = 2$ inches, and $e = \frac{1}{3}$.

We know that $BB_1$, the minor axis, is a line of symmetry in the ellipse; so:

$$BF = BF_1.$$

We also know that the sum of the distances from any point on the ellipse to the two foci is always the same. The sum of the distances from $B$ is $BF + BF_1$; from $A$ it is $AF + AF_1$. So:

$$BF + BF_1 = AF + AF_1$$
or $$2BF = AF + AF_1.$$

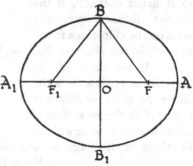

Instead of $AF_1$ we write its equal $AF + OF + OF_1$, or $AF + 2OF$. We now have:
$$2BF = 2AF + 2OF$$
$$= 2OA.$$
So $$BF = OA, \text{ or } a.$$
That is, $BF =$ the semi-major axis.

We can now find a connection between $OF$, the distance of a focus from the centre, and the lengths of the semi-axes. We have got all three lines in the right-angled triangle $BOF$; $BO$ is $b$, and we have found that $BF = a$. The triangle is right-angled, so:

$$BF^2 = OB^2 + OF^2$$

or $\qquad a^2 = b^2 + OF^2.$

So $\qquad OF^2 = a^2 - b^2$

and $\qquad OF = \sqrt{a^2 - b^2}.$

We also know that $\dfrac{OF}{OA} = e$. That is indeed the definition of $e$—the fraction that $OF$ is of $OA$.

So $OF = e \cdot OA$ or $ae$.

Putting the two values of $OF$ together, we have

$$ae = \sqrt{a^2 - b^2}$$

and $\qquad e = \dfrac{\sqrt{a^2 - b^2}}{a}.$

If we know the semi-axes of an ellipse, we can readily find the eccentricity from this formula.

In drawing an ellipse it may be seen that $F_1 P$ increases as the pencil moves from $P$ to $A$; and of course $PF$ decreases. So it arises that $AF_1$ is the greatest distance from any point on the ellipse to a focus, and $AF$ is the shortest distance. If we think of the single focus $F$, then $A$ is the nearest point on the ellipse to it, and $A_1$ is the most distant point. We want to be able to find these lengths when we are told $a$ and $e$.

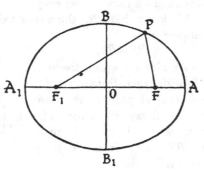

$$A_1 F = A_1 O + OF$$
$$= a + ae$$
$$= a(1 + e)$$

and $\qquad AF = AO - OF$
$$= a - ae$$
$$= a(1 - e).$$

$P$ may represent a planet which is revolving round the sun; the sun is at $F$, and $F_1$ is the empty focus. The point $A$, where the planet is nearest the sun, we have called *perihelion*; the point $A_1$, where it is farthest from the sun, we have called *aphelion*. We now know that

$$\text{perihelion distance} = a(1-e),$$
$$\text{aphelion distance} = a(1+e).$$

In planetary tables it is usual to give the semi-major axis for each orbit, and also the eccentricity; that is, $a$ and $e$. The semi-minor axis is not included because it is seldom wanted, and we can always calculate it if we do want it. Thus:

$$ae = \sqrt{a^2 - b^2},$$
$$a^2 e^2 = a^2 - b^2.$$

So
$$b^2 = a^2 - a^2 e^2$$
$$= a^2(1 - e^2)$$
$$= a(1+e) \times a(1-e)$$
$$= A_1 F \times AF.$$

So $b$ is the geometric mean between $A_1 F$ and $AF_1$ or between $a(1+e)$ and $a(1-e)$. To find $b$ we multiply these two lengths, and take the square root.

It happens occasionally that we know $a$ and $b$. We want to know $e$, and we want to be able to draw the ellipse. A meridian of the earth, for example, is an ellipse. The equatorial radius of the earth is the semi-major axis, and the polar radius is the semi-minor axis.

$$a = 3963\tfrac{1}{3} \text{ miles}, \quad b = 3950 \text{ miles},$$
$$e = \frac{\sqrt{a^2 - b^2}}{a} = \frac{\sqrt{(3963\tfrac{1}{3})^2 - 3950^2}}{3963\tfrac{1}{3}}$$
$$= \cdot 082.$$

So the eccentricity of a meridian is $\cdot 082$, or slightly less than $\frac{1}{12}$.

Suppose we make $a = 5$ inches, then $ae = 5 \times \cdot 082 = \cdot 41$ inch. We draw a line 10 inches long for the major axis, bisect it, and measure out $\cdot 41$ inch on each side of the middle for the two foci. We can then draw the ellipse, which gives the shape of the meridian.

We sometimes want the equation of an ellipse. We know from coordinate geometry that this is:

$$\frac{x^2}{a^2} + \frac{y^2}{b^2} = 1.$$

Thus the equation of a meridian is:

$$\frac{x^2}{(3963\frac{1}{2})^2} + \frac{y^2}{3950^2} = 1.$$

That equation contains full information about the meridian.

If we are told $a$ and $e$ we begin by finding $b$ from the formula:

$$b^2 = a^2(1 - e^2).$$

Now let us apply these ideas to the earth's orbit. It is an ellipse with the sun at one focus, like the orbits of all the other planets.

Planetary tables give the length of the semi-major axis as 92·9 million miles, and the eccentricity as ·01675.

$$a = 92\cdot9 \times 10^6 \text{ miles}, \quad e = \cdot01675.$$

We can write the eccentricity as a fraction:

$$\cdot01675 = \frac{1675}{100,000}$$

$$= \frac{67}{4000}$$

$$= \frac{1}{\frac{4000}{67}}$$

$$= \frac{1}{59\cdot7}.$$

This is very nearly equal to $\frac{1}{60}$, and for most purposes we can take the eccentricity as $\frac{1}{60}$.

$$\tfrac{1}{60} = \cdot01667,$$

so that there is a difference of $1675 - 1667 = 8$ parts in 1675. This is a difference of $\frac{800}{1675}$ per cent = about $\frac{1}{2}$ of one per cent. We can take $e = \frac{1}{60}$, and if we want a more accurate result we can increase the result thus obtained by $\frac{1}{2}$ of one per cent.

Since we know $a$ and $e$ we can construct the earth's orbit to scale. We want a fairly big drawing, so we make the semi-major axis 5 inches long. We draw a line 10 inches long, and bisect it. The distance of each focus from the centre is $ae$.

$$ae = 5 \times \tfrac{1}{60} = \tfrac{1}{12} \text{ inch.}$$

So we fix pins at the foci, each $\frac{1}{12}$ inch out from the centre.

The greatest distance from a focus is:

$$a(1 + e) = 5 \times 1\tfrac{1}{60} = 5\tfrac{1}{12} \text{ inches.}$$

We make a loop of thread $5\frac{1}{12}$ inches long when it is drawn out straight. We put the loop over the pins and draw the orbit.

If we attempt to draw the sun in the diagram we want to know its scale diameter. The actual diameter is 866,000 miles. We have the proportion:

$$92\cdot9 \text{ million miles} = 5 \text{ inches on the scale,}$$

$$1 \text{ million miles} = \frac{5}{92\cdot9} \text{ inch,}$$

$$\cdot866 \text{ million miles} = \frac{5 \times \cdot866}{92\cdot9} \text{ inch}$$

$$= \text{about } \tfrac{1}{20} \text{ inch.}$$

A twentieth of an inch is a rough approximation, but as near as we want for this purpose. Round one focus we draw a small circle of diameter equal to half the tenth of an inch.

The diagram has been reduced from a drawing made in the way just described. The ellipse is so nearly a circle that it is not easy to distinguish the differ-ence by eye. Nevertheless, the difference is great enough to have important conse-quences.

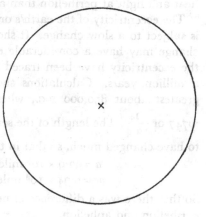

The eccentricity of the earth's orbit is so small that it is often exaggerated in order to show the elongation from the circular form. That is quite legitimate, so long as the exaggeration is under-stood. It is a different matter altogether when the orbit is shown in perspective with a considerable tilt. In the latter case the sun is not at the focus of the ellipse; this may readily be seen by tilting the scale drawing of the earth's orbit; the sun retains its position near the centre.

We are always interested in the greatest and least distances from the sun, because it is in these positions that there is the greatest difference of heating and lighting effect. We want to find $a + ae$ and $a - ae$. We begin by finding $ae$.

$$ae = 92\cdot9 \times \cdot01675 = 1\cdot56 \text{ million miles.}$$

That is the distance the sun is out from the centre of the ellipse.

*Aphelion* (greatest) distance:

$$a + ae = 92\cdot9 + 1\cdot56 = 94\cdot46 \text{ million miles.}$$

*Perihelion* (least) distance:

$$a - ae = 92\cdot9 - 1\cdot56 = 91\cdot34 \text{ million miles.}$$

There is a difference of $1\cdot56 \times 2 = 3\cdot12$, or more than 3 million miles between the greatest and least distances.

The difference in the amounts of light and heat received in the two positions is:

$$\frac{\text{at perihelion}}{\text{at aphelion}} = \frac{(\text{aphelion distance})^2}{(\text{perihelion distance})^2}$$

$$= \frac{94\cdot46^2}{91\cdot34^2}$$

$$= 1\cdot069.$$

That is, the earth as a whole receives about 7 per cent more heat and light at perihelion than at aphelion.

The eccentricity of the earth's orbit is not a fixed quantity, but is subject to a slow change. It should be clear now that such a change may have a considerable effect on climate. Changes in the eccentricity have been traced back for periods extending to a million years. Calculations show that the eccentricity was greatest about 850,000 B.C., when it reached a maximum of $\cdot0747$ or $\dfrac{1}{13\cdot4}$. The length of the semi-major axis does not appear to have changed much, so that in that far distant epoch we had:

$$a = 92\cdot9 \times 10^6 \text{ miles,} \quad e = \cdot0747.$$
$$ae = 6\cdot94 \times 10^6 \text{ miles.}$$

So that there was a difference of nearly 14 million miles between perihelion and aphelion.

*Aphelion*:     $a + ae = 92\cdot9 + 6\cdot94 = 99\cdot84$ million miles.

*Perihelion*:     $a - ae = 92\cdot9 - 6\cdot94 = 85\cdot96$ million miles.

So that in 850,000 B.C.:

$$\frac{\text{heat received at perihelion}}{\text{heat received at aphelion}} = \frac{99\cdot84^2}{85\cdot96^2}$$

$$= 1\cdot349.$$

That is to say, the earth received nearly 35 per cent more heat at perihelion than at aphelion. That difference was sufficient to produce very hot summers and very cold winters in one hemisphere; and cool summers and mild winters in the other. These extreme differences would occur in the periods nearest to 850,000 B.C. when the solstices were near perihelion and aphelion.

Let us look at some of the other planets. The most nearly circular orbit is that of Venus. The eccentricity is only ·00682 = about $\frac{1}{147}$. For Venus we have:

$$a = 67·2 \text{ million miles}, \quad e = ·00682.$$
$$ae = 67·2 \times ·00682 = ·46 \text{ million miles}.$$

Aphelion distance:

$$a + ae = 67·2 + ·46 = 67·66 \text{ million miles}.$$

Perihelion distance:

$$a - ae = 67·2 - ·46 = 66·74 \text{ million miles}.$$

The difference in the amounts of heat received is:

$$\frac{67·66^2}{66·74^2} = 1·028.$$

So that there is a difference of only about $2\frac{3}{4}$ per cent between perihelion and aphelion, as compared with 7 per cent for the earth.

Far and away the most elongated orbit amongst those of the inner planets is that of Mercury, the planet nearest the sun. The eccentricity of this orbit is a little over $\frac{1}{5}$.

$$a = 36·0 \times 10^6 \text{ miles}, \quad e = ·2056.$$
$$ae = 36·0 \times ·2056 = 7·4 \text{ million miles}.$$

Aphelion distance:

$$a + ae = 36·0 + 7·4$$
$$= 43·4 \text{ million miles}.$$

Perihelion distance:

$$a - ae = 36·0 - 7·4$$
$$= 28·6 \text{ million miles}.$$

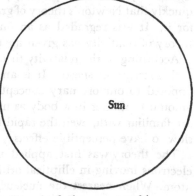

The diagram shows the orbit of Mercury drawn to scale. The ellipticity can readily be seen. If we take as before 10 inches to represent the major axis, we have:

$$a = 5 \text{ inches}, \quad ae = 5 \times ·2056 = 1 \text{ inch (near enough)}.$$

To find the foci we measure out 1 inch on each side of the centre.

To find the width of the sun in the diagram we have:

$$36 \text{ million miles} = 5 \text{ inches,}$$

$$\cdot 866 \text{ million miles} = \frac{5 \times \cdot 866}{36} \text{ inch}$$

$$= \cdot 12 \text{ inch,}$$

or about an eighth of an inch. That is the width of the sun in the diagram.

The great eccentricity of the orbit of Mercury makes a considerable difference in the amount of heat and light received at perihelion and aphelion. The difference is:

$$\frac{43 \cdot 4^2}{28 \cdot 6^2} = 2 \cdot 30.$$

So that nearly $2\frac{1}{3}$ times as much light and heat are received at perihelion as at aphelion.

The great eccentricity also introduces a considerable difference in the speed of the planet at different parts of its orbit. We found in the case of the earth that the perihelion speed is $1 \cdot 034$ times the aphelion speed. Using the same argument, we find for Mercury:

$$\frac{43 \cdot 4}{28 \cdot 6} = 1 \cdot 52.$$

So that the perihelion speed is 52 per cent greater than the aphelion speed.

Mercury, like other planets, precesses. In fact it precesses so quickly that Newton's theory of gravitation was unable to account for it. It was regraded as an unexplained anomaly. Einstein's theory of relativity has given the desired explanation.

According to the relativity theory the mass of a body increases with increase of speed. It is an extraordinary idea, and quite opposed to our ordinary conceptions; we commonly regard the amount of matter in a body as fixed. The kind of movement we are familiar with, even the rapid flight of an aeroplane, is far too slow to have perceptible effects.

The theory was first applied to the movements of very fast electrons moving in elliptical orbits round a heavy nucleus in an atom. When nearest the nucleus, the speed was increased, and the mass of the atom increased. The heavy electron is less affected by the attraction of the nucleus than when it is lighter; it swings out from the nucleus, and so the orbit is caused to roll in a direction retrograde to the direction in which the electron is revolving.

Mercury is the planet nearest to the sun, and therefore most influenced by the attraction of the sun. Also its orbit has a great eccentricity, so that there is a great increase in speed when the planet is nearest to the sun. The increase in mass in these conditions is sufficient to account for the rolling of the orbit that is seen as precession.

The surface of Mars is a matter of great interest, because in some ways Mars is the planet most like the earth. It has rather more than half the diameter of the earth. It is about half as far away again from the sun as the earth is, so that it receives something like half as much light and heat as fall on similar areas of the earth. It appears to have an atmosphere. The axis is slightly more inclined than the earth's axis, so that it has the same kind of seasonal changes. Opportunities for studying Mars at comparatively close quarters are therefore eagerly looked for.

We want the perihelion and aphelion distances of Mars. We have:
$$a = 141 \cdot 6 \times 10^6 \text{ miles}, \quad e = \cdot 0933,$$
$$ae = 141 \cdot 6 \times \cdot 0933 \times 10^6 \text{ miles} = 13 \cdot 2 \text{ million miles}.$$

Aphelion distance:
$$a + ae = 141 \cdot 6 + 13 \cdot 2 = 154 \cdot 8 \text{ million miles}.$$

Perihelion distance:
$$a - ae = 141 \cdot 6 - 13 \cdot 2 = 128 \cdot 4 \text{ million miles}.$$

The greater eccentricity of Mars' orbit makes it more important to have Mars close in than to have the earth far out toward Mars. Unfortunately there is a considerable difference in the longitudes of Mars perihelion and the earth's aphelion; the difference is over 50°, so that when Mars is at its nearest point to the sun the nearest point on the earth's axis is farther away than if this were aphelion for the earth.

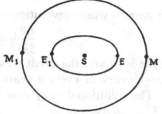

The diagram shows that the earth and Mars are comparatively close together when the earth is at $E$ and Mars at $M$, or the earth at $E_1$ and Mars at $M_1$; that is, when they are in line with the sun, and on the same side of it. The greatest distances apart are $E$ and $M_1$, and $E_1$ and $M$; that is, when the earth and Mars are in line with the sun, and on opposite sides of it.

Perihelion distance of Mars — aphelion distance of earth

$$= 128 \cdot 4 - 94 \cdot 46 = 33 \cdot 94 \text{ million miles.}$$

That is the nearest possible approach of Mars. As we have seen it does not actually approach the earth quite so closely.

Aphelion distance of Mars + aphelion distance of earth

$$= 154 \cdot 8 + 94 \cdot 46 = 249 \cdot 26 \text{ million miles.}$$

That is the greatest possible distance apart of the earth and Mars. And again, Mars is never quite so far distant as that from the earth.

The difference between the two distances is striking. The greater is $\frac{249}{34} =$ about 7 times the less. The difference in the area (and brightness) of Mars is therefore $\frac{1}{7^2} =$ about $\frac{1}{50}$. Or, Mars is about 50 times as bright when it is nearest, as it is when it is farthest away.

We want to find at what intervals the earth and Mars come together at the same side of the sun; or at what intervals the earth comes between Mars and the sun. The earth goes round the sun in $365\frac{1}{4}$ days, and Mars in 687 days.

Suppose the planets start off at a position of near approach (the earth between the sun and Mars). In $365\frac{1}{4}$ days the earth has performed a whole revolution and Mars $\dfrac{365\frac{1}{4}}{687}$ of a revolution. In a year the earth has gained:

$$1 - \frac{365\frac{1}{4}}{687} = \frac{687 - 365\frac{1}{4}}{687}$$

$$= \frac{321\frac{3}{4}}{687} \text{ of a revolution.}$$

It gains a whole revolution in:

$$\frac{687}{321\frac{3}{4}} = 2 \cdot 135 \text{ years.}$$

That is to say, the earth passes between the sun and Mars every $2 \cdot 135$ years, or every 2 years $49 \cdot 3$ days.

The additional $\cdot 135$ year carries the earth round the sun in $\dfrac{1}{\cdot 135} = 7 \cdot 4$ periods.

We want an integral number of periods, so that Mars and the earth may be in the same places in the orbits as at the beginning. The smallest multiple of 7·4 which gives an integer is:

$$7·4 \times 5 = 37.$$

37 periods of 2·135 years = 79 years. And this is the number of years in which the sun, the earth, and Mars return to the same relative positions. In 79 years the earth goes round the sun 79 times, and Mars 42 times.

$$365\tfrac{1}{4} \times 79 = 28,854\tfrac{3}{4} \text{ days,}$$
$$687 \times 42 = 28,854 \text{ days,}$$

so that the discrepancy is small.

After being at their points of closest approach, it is 79 years before the earth and Mars return to these points at the same time.

The most recently discovered planet, Pluto, has a very elliptical orbit, more elliptical even than that of Mercury. The semi-major axis is given as 39·52 astronomical units:

$$= 39·52 \times 92·9 \times 10^6 \text{ miles} = 3671 \text{ million miles.}$$

The eccentricity is ·2486.

$$ae = 3671 \times 10^6 \times ·2486 = 912·5 \text{ million miles.}$$

Perihelion distance $= a - ae = 2758·5$ million miles.
Aphelion distance $= a + ae = 4583·5$ million miles.

The diagram shows the orbit of Pluto drawn to scale. It is an interesting point about this orbit that the perihelion distance of Pluto is less than the semi-major axis of the orbit of Neptune, so that part of the orbit of Pluto lies inside the orbit of Neptune. One cannot help wondering what will happen when the two planets come together in the intersecting parts of their orbits, and why they have not come together already, some time in the 3000 million years or more of the past history of the solar system.

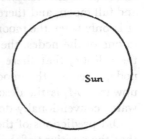

The orbit of each of the planets lies in a plane, just as if it were drawn on an enormous flat sheet of paper. These planes are inclined to each other at small angles; so that the whole solar system is contained in a cylinder of comparatively small height, whose width is the outside width of the orbit of Pluto.

The plane of the earth's orbit is used as a zero plane; it is called the *ecliptic*, because eclipses occur when the moon is in this plane. The plane of the orbit of Mercury is inclined at 7° to the ecliptic and the planes of the other planets at lesser angles. For this reason the planets are never seen outside the zodiac, the band of constellations through which the sun makes its yearly widdershins circuit of the heavens. The zodiac extends 7° north and south of the ecliptic so as to include the orbit of Venus. The orbit of the moon is inclined at just over 5° to the ecliptic, so that the moon also appears to circle through the zodiac. Month after month the moon may be seen in one after another of the constellations of the zodiac, and it may be seen to pass fairly close to Venus, Mars and Jupiter.

If the plane of the moon's orbit were coincident with the ecliptic, then at every new moon the moon would come exactly between the earth and the sun, and there would be a total eclipse of the sun. At every full moon the earth would come between the sun and the moon, and there would be a total eclipse of the moon. As things are, it usually happens that the moon is north or south of the ecliptic at new moon and full moon, and there is no eclipse. It is only when the moon is at or close 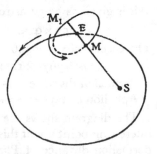 to one of the nodes, the points where the orbit of the moon cuts the ecliptic, that there is an eclipse. $M$ is called the ascending node, because the moon is moving northward, conventionally upward. $M_1$ is the descending node; the moon is moving southward, conventionally down.

Old predictions of the times of eclipses were based on the fact that the positions of the sun and moon relative to the earth recur at certain intervals. The average interval between new moon and new moon (the synodical month, between one ascending node and the next) is 29 days $12\frac{3}{4}$ hours.

$$29 \text{ days } 12\frac{3}{4} \text{ hours} = 2835 \text{ quarter hours,}$$
$$365\frac{1}{4} \text{ days} = 35,064 \text{ quarter hours,}$$
$$\frac{35,064}{2835} = 12 \text{ and } 1044 \text{ over.}$$

That is, in a Julian year of $365\frac{1}{4}$ days, there are 12 new moons, and 1044 quarter hours toward the 2835 quarter hours which make up another synodical month. We want to find a multiple of 1044 which is an exact, or almost exact, multiple of 2835. We keep adding multiples of 1044, and subtracting 2835 when possible, until we come close to a remainder of 2835 or 0.

$$
\begin{array}{ll}
1044 \times 3 & = 3132 \\
\text{Subtract } 2835 & 297 \\
\text{Add } 1044 \times 3 & 3429 \\
\text{Subtract } 2835 & 594 \\
\text{Add } 1044 \times 3 & 3726 \\
\text{Subtract } 2835 & 891 \\
\text{Add } 1044 \times 3 & 4023 \\
\text{Subtract } 2835 & 1188 \\
\text{Add } 1044 \times 2 & 3276 \\
\text{Subtract } 2835 & 441 \\
\text{Add } 1044 \times 3 & 3573 \\
\text{Subtract } 2835 & 738 \\
\text{Add } 1044 \times 2 & 2826 \\
\end{array}
$$

And 2826 is not greatly different from 2835.

Hence $1044 \times 19 = 2835 \times 7$ (very nearly).

So that in a period of 19 years there are 7 extra new moons, or $19 \times 12 + 7 = 235$ altogether. In a period of 19 years there is an almost exact number of synodical months. At the end of the period the moon returns to the same position, relative to the earth and sun, as at the beginning of the period. Thus we have a repetition of eclipses and other phenomena connected with the positions of sun, earth and moon every 19 years.

The period of 19 years is called the Metonic cycle. It is so named after Meton, an Athenian astronomer who discovered it in the fifth century B.C. Successful predictions of eclipses were made on the basis of the Metonic cycle, but it began to appear that the period was a very little too long. Eclipses came a little before the predicted times at first, and then more and more before the predicted times. A more exact cycle was discovered by Callipus in the following century. The Callipic cycle is a day less than four Metonic cycles, that is 76 years all but a day.

# CHAPTER 7

## *Proportions in the Stars*

IT REALLY is extraordinary how numbers seem to acquire wings. The only place where we can make an actual measurement of length is on the earth itself, and there we measure laboriously, yard by yard, a length of a few miles. But having, with care and patience, got that original measurement, firm based upon the earth, the wings of arithmetic can carry it over a whole country, over a continent and the oblate earth itself; then they can carry it over the far-stretched solar system, and finally to the bounds of the visible universe, and perhaps beyond.

In 1609 Kepler completed a long investigation of the orbit of Mars. This planet had attracted great attention because of the considerable eccentricity of its orbit, not far short of one-tenth. From a study of this orbit Kepler was able to announce two of the laws that are known by his name:

1. The planets move round the sun in ellipses. The sun is at one focus of each orbit.

2. The line joining the sun to a planet sweeps over equal areas of the orbit in equal times.

Kepler's Third Law was published ten years later, in 1619. This law is familiarly, and rather awkwardly, referred to as 'the sesquiplicate ratio between the planetary periods and distances'. Sesquiplicate is merely a way of saying '$1\frac{1}{2}$ times', or 'half as much again'. The planetary periods are the times taken by the planets to move round the sun. The distances are the semi-major axes of the orbits.

According to Kepler's Third Law: the square of the period is proportional to the cube of the distance.

$$\text{period}^2 = c \ \text{distance}^3$$
$$\text{or period} = k \ \text{distance}^{\frac{3}{2}}, \qquad \text{or } k \ \text{distance}^{1\frac{1}{2}}$$

That is where the 'sesquiplicate' comes in.

The constant $c$ or $k$ (whichever we use) is the same for every planet. If we are considering a system of satellites revolving round a planet, we should still have a constant for all the satellites, but it would be a different constant from the one for the planets.

Measurements of time can be made with great exactness, and so also can angular measurements. Clocks are made which vary from exact time by less than a tenth of a second in a day. Even that variation is only a tenth of a second in 86,400 seconds, or 1 part in 864,000.

Kepler's Third Law enables us to relate the very exact measurements of planetary periods to the proportions of their distances from the sun. We have the equation:

$$p^2 = cd^3.$$

By taking the earth's time (the sidereal year) and the earth's distance as units, we have for the earth:

$$1^2 = c \times 1^3,$$

so $c = 1$, and so long as we keep these units, $c$ remains 1. The period of Mars is known to be:

$$686 \cdot 9797 \text{ days} = \frac{686 \cdot 9797}{365 \cdot 2564} \text{ sidereal years}$$

$$= 1 \cdot 880815 \text{ sidereal years}.$$

The equation for Mars is:

$$1 \cdot 880815^2 = d^3.$$

$$d = 1 \cdot 880815^{\frac{2}{3}}$$

$$= 1 \cdot 52369.$$

(We can work that out, either by using logarithms, or by squaring $1 \cdot 880815$ and then finding the cube root.)

It should be clear now that we have got an extremely accurate scale model of the solar system, using the semi-major axis of the earth's orbit (the astronomical unit) as unit.

In passing we may note that it is possible to use Kepler's Third Law in the reverse direction to find the periodic time of a planet. The greatest and least distances of Pluto, the newly discovered planet beyond Neptune (or usually beyond Neptune), are given as 49·49 and

29·55 astronomical units. Thus the semi-major axis of Pluto's orbit is:

$$\frac{49\cdot49 + 29\cdot55}{2} = 39\cdot5 \text{ astronomical units.}$$

$$\text{period} = \text{distance}^{\frac{3}{2}}$$
$$= 39\cdot5^{\frac{3}{2}} \text{ years}$$
$$= 248 \text{ years.}$$

Another point that should be clear is this: if we can measure a single line in the solar system with any degree of accuracy, we can spread this measurement over the whole scale model and so find all the lengths in miles. A single measurement gives the scale to which the model is drawn. The degree of accuracy of the single measurement gives the degree of accuracy for all measurements in miles throughout the solar system.

That is why it is important to have one good measurement of length.

From each end of his measured base line the surveyor turns a telescope on a distant point. He measures the angles between the base line and the distant point, and he uses these angles to calculate the distance of the point from each end of the base line. The method of the astronomer is the same in principle. The first difficulty is to find an adequate base line. The longest line available is the diameter of the earth, or at least a large fraction of it.

We can have two observatories a considerable distance apart on the earth's surface, and we can readily calculate their distance apart in a straight line. At the same time both telescopes are turned to a spot on the sun's surface; and the angles at which the telescopes are placed is noted. The slight difference in inclination is found, and from this the distance of the sun can be calculated.

As soon as attempts were made to measure the angles, it was found that there was a great difficulty in fixing an exact spot for observation. This difficulty will be appreciated when it is realised that the observers were trying to measure the angle to the thousandth part of a second. We have to imagine a degree divided into 60 minutes, then each of these divisions into 60 seconds, and the second further divided into 1000 parts. The whole angle to be measured was less than 20″, so a very small discrepancy was enough to vitiate the measurement.

The change of angle when an object is looked at from two points of view is called *parallax*. There is one example of parallax we are all familiar with—parallax on a clock face. At the half hour we stand right in front of the clock, and we see the minute hand exactly over the 6. Move to the right and we see the minute hand a little after the 6; that is, it appears to show a little after the half hour. Move to the left, and the minute hand appears to show a little before the half hour.

I read a story in which one of the characters was said to have removed the minute hand from a clock, in order that witnesses might mistake the shadow of the hour hand for the minute hand, and so cause confusion about the time. The story goes on: 'The witnesses differ according to the direction in which they saw the shadow fall. Professor Ingram, sitting to the extreme right, saw the shadow fall at one minute to twelve. Miss Wills, sitting in the centre, saw it dead on midnight. This film, taken from the extreme left, shows it at one minute past twelve.'

With all due disrespect to both villain and detective, they were both wrong. The villain would have got a much larger parallax by leaving the minute hand in position; as it was, the only parallax he had was the much smaller parallax of the hour hand, smaller because the hour hand is closer to the clock face. He also had to gamble on the fact that his audience (including Professor Ingram) were so supremely ignorant or unintelligent as not to allow for parallax. The detective was wrong in ascribing parallax to the shadow. The shadow is actually on the clock face, and therefore shows no parallax.

Occasionally Venus comes exactly between the earth and the sun. (Usually it is too far north or south, on the occasions when it passes between the earth and the sun, to be seen with the sun as a background.) When there actually is a transit of Venus, there is an opportunity, not to be missed, of measuring its parallax.

*A* and *B* are two observatories wide apart on the earth's surface. From *A*, Venus is seen at *C*, and from *B* it is seen at *D*. The distance *AB* can be calculated, and from this the distance of Venus can be found. From this again the parallax of the sun can be found.

It seemed a great opportunity, but the result was disappointing. Venus has too great a disk, and the sun is too brilliant a background, to permit of extreme accuracy in the measurements. Better results have been obtained by observation of the small asteroid, Eros. The Astronomer Royal found 8·79″ for the sun's mean equatorial parallax. The distance of the sun is therefore the equatorial radius of the earth, 3963·3 miles, divided by the tangent of 8·79″. The calculation is as follows:

$$180° = 3·14159 \text{ radians,}$$
$$1″ = 3·14159 \div 180 \div 3600 \text{ radian}$$
$$= ·0000048481 \text{ radian,}$$
$$8·79″ = ·0000048481 \times 8·79 \text{ radian}$$
$$= ·0000426148 \text{ radian.}$$

So that     $\tan 8·79″ = ·0000426148.$

The distance of the sun $= \dfrac{3963·3}{·0000426148}$ miles

$$= 93·0 \text{ million miles.}$$

The distance of the sun is often given as 92·9 million miles.

We will now see how the measurement of the sun's distance can be used to find its radius. Measurement shows that at its mean distance the sun has an angular radius of 16′ 1·18″ at its equator.

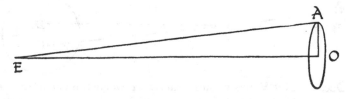

In this diagram angle *AEO* is the angular radius of the sun. We can take *AE* as the distance of the sun because there is little difference between *AE* and *OE*.

$$\sin AEO = \frac{AO}{AE} = \frac{\text{radius of sun}}{\text{distance of sun}}.$$

Hence, radius of sun = distance of sun × sin 16′ 1·18″.

16′ 1·18″ is still within the region where the sine can be taken as equal to the circular measure of the angle.

$$16' \; 1{\cdot}18'' = 961{\cdot}18''$$
$$= {\cdot}0000048481 \times 961{\cdot}18 \text{ radian}$$
$$= {\cdot}0046599 \text{ radian},$$
$$\sin 16' \; 1{\cdot}18'' = {\cdot}0046599.$$

The equatorial radius of the sun:

$$= 92{\cdot}9 \times 10^6 \times {\cdot}0046599 \text{ miles}$$
$$= 432{,}900 \text{ miles, or } 433{,}000 \text{ miles}.$$

If we compare the sun's radius with that of the earth, we find it is:

$$\frac{433{,}000}{3963{\cdot}3} = 109 \text{ times as great}.$$

We can consider the sun as having very nearly the same shape as the earth; so its area is:

$$109^2 = 11{,}881, \text{ or about } 12{,}000 \text{ times as great}.$$

The volume is:

$$109^3 = 1{,}295{,}000, \text{ or about } 1{,}300{,}000 \text{ times as great}.$$

The volume of the sun is sometimes given as a million times that of the earth, but this is 30 parts in 130, or a 23 per cent understatement.

The problem of finding the distance of a star is not different in kind from that of finding the distance of a planet, though the necessary measurements are much more difficult to make. The original trouble was to find a base line long enough. A half rotation of the earth at the equator gives a base line of 8000 miles. This proved quite inadequate; there seemed to be no parallax at all. We know now that 40 billion miles is a comparatively small star distance, so that the sine of the angle to be measured was:

$$\frac{\text{length of base line}}{\text{distance of star}} = \frac{8000 \text{ miles}}{40 \times 10^{12} \text{ miles}} = \frac{2}{10^{10}}.$$

The angle is: $\dfrac{2}{10^{10}}$ radians $= \dfrac{2 \times 10^{-10}}{\cdot000004848}$ second

$$= \dfrac{1}{24240} \text{ second.}$$

Or about the twenty-four thousandth part of a second. It is no wonder the angle could not be measured.

A suitable base line was found by making measurements at opposite ends of the earth's orbit. This gave a base line of about 186 million miles, which is about 23,000 times as great as the 8000 miles available on the earth. Multiplication by 23,000 would change $\frac{1}{24240}$ of a second into about a second, so that a difficult but possible measurement was suggested. The interval of six months between observations introduced the difficulty of slight shifts in the actual positions of the stars, quite apart from the apparent shift due to parallax. The two have to be disentangled before the star distance can be calculated.

Star distances are sometimes given in light-years, sometimes in parsecs, and occasionally in billions of miles. A light-year is the distance travelled by light in a year; it is about 5·9 billion miles.

A *parsec* is the distance of a star which has a parallax of 1″. We can find this distance also in billions of miles.

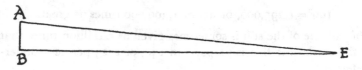

In the diagram $AB$ is the length of the base line, 92·9 million miles. $AEB$ is the parallax; actually this should be 1″.

$$\sin 1'' = \frac{AB}{AE} = \frac{92 \cdot 9 \times 10^6 \text{ miles}}{1 \text{ parsec}}.$$

$$1 \text{ parsec} = \frac{92 \cdot 9 \times 10^6}{\sin 1''} \text{ miles}$$

$$= \frac{92 \cdot 9 \times 10^6}{\cdot000004848} \text{ miles}$$

$$= 19 \cdot 2 \text{ billion miles.}$$

There is an advantage in using the parsec as a unit; it is closely

connected with the parallax. We have seen that the distance of
a star is given by:

$$\frac{92 \cdot 9 \times 10^6 \text{ miles}}{\text{sine of parallax}} = \frac{92 \cdot 9 \times 10^6 \text{ miles}}{\cdot000004848 \times \text{parallax in seconds}}$$

$$= \frac{1 \text{ parsec}}{\text{parallax in seconds}}.$$

That is, to find the number of parsecs we divide by the parallax
in seconds. This is only true of course for angles so small that the
sine is equal to the circular measure of the angle.

The parallax of Alpha Centauri, the nearest star to be measured,
is ·75″. Its distance is therefore

$$\frac{1}{\cdot75} \text{ parsecs} = 1 \cdot 33 \text{ parsecs}.$$

The parallax of Rigel, the bright star at the heel of Orion, is
given as ·006″. Its distance is:

$$\frac{1}{\cdot006} = \frac{1000}{6} = 167 \text{ parsecs}.$$

If we want the distance in miles, it is:

$$167 \times 19 \cdot 2 \text{ billion miles}$$
$$= 3200 \text{ billion miles}.$$

In light-years this is:

$$\frac{3200}{5 \cdot 9} = \text{about } 540 \text{ light-years}.$$

We have seen how measurements can be extended from short
base lines on the earth to such improbable lengths as thousands
of billions of miles. Let us now look at some comparisons within
the solar system.

It is a small oddity of arithmetic that the effect of distance on
apparent length is less well known than its effect on apparent
area. We all know about the inverse square rule: that apparent
areas are inversely proportional to the square of the distance;
that if we remove an area to twice the distance it appears to have
only a quarter of the area. We are less well aware of the fact that
apparent lengths are inversely proportional to the distances; that
if we double the distance we halve the apparent length.

To an observer with his eyes at $A$, the lengths $BB_1$, $CC_1$, $DD_1$, $EE_1$ all appear the same length. If they were rods, the rod $BB_1$ would exactly cover any of the others. The lengths $AB$, $BC$, $CD$ and $DE$ have all been made equal. So $CC_1$ is twice $BB_1$, and its

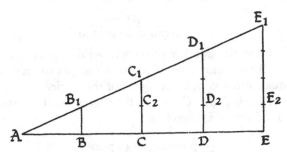

apparent height is halved by doubling its distance. $DD_1$ is three times $BB_1$, so its apparent height is divided by three when we multiply the distance by three. $EE_1$ is four times $BB_1$, so its apparent height is reduced to a quarter when we multiply the distance by four.

This is the idea that is used in the ring sight. An observer at $A$ looks through the ring at $B$ and sees the aeroplane $C$, bearing down on him and just filling the ring. If he knows the width of

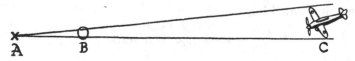

the plane he can find its distance. Suppose the length $AB$ is two feet, and the width of the ring one inch. The observer recognises the plane as having a width of 120 feet. We have the proportion:

$$\frac{AB}{1 \text{ inch}} = \frac{AC}{120 \text{ feet}}$$

or

$$\frac{2 \text{ feet}}{1 \text{ inch}} = \frac{24 \text{ inches}}{1 \text{ inch}} = \frac{AC}{120 \text{ feet}}.$$

$$AC = 24 \times 120 \text{ feet}$$
$$= 960 \text{ yards}.$$

That is, the observer need only multiply the recognised width of the pursuing plane in feet by 8, to find its distance in yards.

If the pursuing plane fills half the ring its distance is twice as great: $960 \times 2 = 1920$ yards. If it fills $\frac{2}{3}$ of the ring we have to multiply by $\frac{3}{2}$; the distance is $960 \times \frac{3}{2} = 1440$ yards. And so on for any other fraction. We multiply by the reciprocal of the fraction, that is by one over the fraction, or the fraction inverted.

By the same simple method we can compare the width of the sun as seen from any planet with its width as seen from the earth. A table gives the distances of the planets using the earth's distance as unit. Thus the distance of Mercury (the semi-major axis of the orbit) is given as ·387 times the earth's distance.

So the width of the sun as seen from Mercury:

$$= \frac{1}{\cdot 387} = 2 \cdot 584 \text{ times its width as seen from the earth.}$$

The apparent widths of the sun as seen from the other planets are found in the same way. They are:

Venus: $\dfrac{1}{\cdot 723} = 1 \cdot 383$, or about $1\frac{1}{3}$ times as wide as when seen from the earth.

Mars: $\dfrac{1}{1 \cdot 52} =$ about $\frac{2}{3}$ as wide. (Half as wide as from Venus, because Mars is more than twice as far off as Venus.)

Jupiter: $\dfrac{1}{5 \cdot 2} =$ less than $\frac{1}{5}$ as wide.

Saturn: $\dfrac{1}{9 \cdot 54} =$ about $\frac{2}{19}$, or less than $\frac{1}{9}$ as wide.

Uranus: $\dfrac{1}{19 \cdot 2} =$ less than $\frac{1}{19}$ as wide. (Half as wide as from Saturn, because Uranus is about twice as far off as Saturn.)

Neptune: $\dfrac{1}{30} =$ as wide.

We can use these numbers to represent the size of the sun as it would appear from each of the planets. We begin by choosing a standard size to represent the sun as seen from the earth. We may have a circle of white paper, 10 inches in diameter; and we want to know at what distance to place this circle so that it may represent the apparent size of the sun as seen from the earth. It is a slight variation of the problem of the ring sight.

In the diagram $AB$ is the diameter of the ten-inch circle which we are using as a standard. $CD$ is the diameter of the sun, and $OC$ its distance from the earth. $x$ is the distance we want to find.

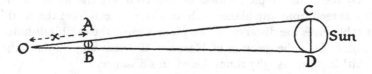

The two triangles $OAB$ and $OCD$ are similar, so $x$ is the same fraction of $OC$ that $AB$ is of $CD$.

$$\frac{x}{OC} = \frac{AB}{CD},$$

$$\frac{x}{\text{distance of sun}} = \frac{10 \text{ inches}}{\text{diameter of sun}},$$

$$\frac{x}{92,900,000 \text{ miles}} = \frac{10 \text{ inches}}{866,000 \text{ miles}}.$$

(We need not reduce the sun length to inches, because the factors we should use to do so are the same on both sides of the equation, and cancel out.)

$$x = \frac{10 \times 92 \cdot 9 \times 10^6}{866,000} \text{ inches}$$

$$= 1073 \text{ inches}$$

$$= \text{about 30 yards.}$$

That is to say, if we fix on the wall a circle of paper 10 inches across, and stand back 30 yards from it, we shall see it the apparent size of the sun. The truth of this is apparent when we compare the two. Such a circle would exactly cover the sun if it were held between the eyes and the sun at a distance of 30 yards.

We can now cut out circles of white paper to represent the sun as it would appear from each of the planets. The circles have to be viewed from the standard distance of 30 yards. To find the diameters of the circles in inches, we have only to multiply the numbers already found by 10. The diameters are:

Mercury:   25·8 inches.         Venus:     13·8 inches.
Mars       6½ inches.           Jupiter:   2 inches.
Saturn:    rather more than an inch.
Uranus:    ½ inch.              Neptune:   ⅓ inch.

We can of course increase or decrease the scale just as much as we like, provided we multiply or divide all the lengths, including the distances from which the circles are viewed, by the same number. If we were to reduce the scale to a tenth of that described, the sun as seen from the earth would be represented by a circle one inch across, and it would have to be viewed from a point $\frac{30}{10} = 3$ yards away. We may note in passing that a halfpenny is one inch across; it would exactly cover the sun if it were held at a distance of 3 yards. A penny is 1⅓ inches across; it would exactly cover the sun if it were held at a distance of $\frac{30 \times 1\frac{1}{3}}{10} = 4$ yards.

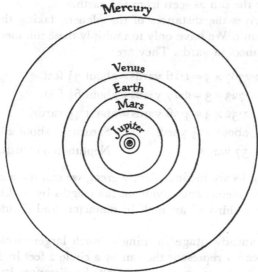

The diagram shows the size of the sun as seen from the different planets. The size of the sun as seen from the earth is a circle 1 inch in diameter, so the diagram should be viewed from a distance of 3 yards.

When we look at these circles we have to remember that the white paper disks only tell a part of the truth. They show the apparent size of the sun accurately enough. What is missing is the intolerable brillance that hurts earthly eyes when we inadvertently look at the sun, and that raises even the inadequate Neptunian sun high out of the catalogue of first magnitude stars. Nevertheless, the appearance of the small circle that represents the Neptunian sun, as compared with the inch circle that represents the sun as we see it, does give an impressive idea of how inadequate that sun is.

There is a rather different way of representing the sun as seen from the various planets; this method has some advantages of its own. We can use the same circle to represent the sun, and we can stand back from it distances proportional to the distances of the planets from the sun. We had better not make the circle too big, or the distances will be unduly long. We can start with a one-inch circle. We know that it has to be placed at a distance of 3 yards to represent the sun as seen from the earth.

A table gives the distances of the planets, taking the earth's distance as unit. We have only to multiply these numbers by 3 to get the distances in yards. They are:

Mercury: ·387 × 3 = 1·16 yards = about 3½ feet.

Venus:   ·723 × 3 = 2·17 yards = about 6½ feet.

Mars:    1·52 × 3 = 4·56 yards = about 4½ yards.

Jupiter: about 15½ yards.        Saturn:    about 28½ yards.

Uranus:  57 yards.               Neptune: 90 yards.

If the lengths are inconveniently great, we can use a circle half an inch in diameter, and divide all the lengths by 2. Or we can use a circle a third of an inch in diameter, and divide all the lengths by 3.

There is an advantage in using a much larger circle for the sun. Suppose we represent the sun by a circle 2 feet in diameter, the scale is 24 times as great, and all the distances have to be multiplied by 24. They range from 28 yards for Mercury and 72 yards for the earth, up to about a mile and a quarter for Neptune. We can cut out a 2-foot circle, from a sheet of newspaper perhaps, and fix it upright so that we can view it from varying distances.

This model is interesting because we can use it to show the effect of moving up close to the sun. Suppose the sun were to move up until its nearest point was at the distance of the moon—about a quarter of a million miles away. We have the proportion:

$$866,000 \text{ miles} = 24 \text{ inches on the scale,}$$

$$250,000 \text{ miles} = \frac{24 \times 250}{866} \text{ inches}$$

$$= 7 \text{ inches.}$$

So that if we bring the eyes up to a distance of 7 inches from the centre of the 2-foot circle which represents the sun, we get the effect (minus the intolerable brilliance and other devastating effects) of the sun as it would appear at the distance of the moon. Viewed from this distance the sun would almost completely fill the sky. Not that we should have long to look at it.

Coming, in pale imagination, thus close to the sun, suggests an interesting comparison between the circumference of the earth, the distance of the moon, and the size of the sun.

The circumference of the earth is a little less than 25,000 miles. The distance of the moon is about 239,000 miles. So that a little more than $9\frac{1}{2}$ days' rotation of the earth, if it were rolling toward the moon at the equatorial speed of a thousand miles an hour, would take it to the moon.

We have some sort of feeling about the enormous distance of the moon. It is something of a shock to realize that if the earth were at the centre of the sun, there would be ample room within the sun for the moon's orbit. The sun's radius is 433,000 miles, so that the moon's orbit would stretch out rather more than halfway toward the circumference.

If we try to imagine the sphere on which the orbit of the moon lies, we have to think of it stretching out in all directions about the earth. As though the whole sky on all sides of the earth were covered with moons, and the vast space thus enclosed filled with hot material.

But even this imitation sun, vast as it would be, is not a very large fraction of the real sun. In comparing the volumes of the two we have to cube similar lengths, so we have:

$$\frac{\text{volume of sun}}{\text{volume of sphere with radius equal to the moon's distance}}$$
$$= \frac{(\text{radius of sun})^3}{(\text{distance of moon})^3}$$
$$= \frac{433^3}{239^3}$$
$$= 6.$$

So that vast sphere has but a sixth part of the volume of the still vaster sun.

We have found a comparison for the width of the sun as seen from the various planets. To compare the area of the sun as seen from the planets we have merely to square these numbers. We found, for example, that the width of the sun as seen from Mercury is 2·584 times its width as seen from the earth. Its apparent area is therefore:

2·584² = 6·667, or 6⅔ times as great as when seen from the earth.

The area of the sun, as seen from each of the other planets is:

Venus:  1·383² = 1·91 or 1$\frac{9}{10}$ times as great as when seen from the earth.

Mars:  $\frac{1}{1\cdot52^2} = \frac{1}{2\cdot31} = \frac{4}{9}$ times as great.

Jupiter:  $\frac{1}{5\cdot2^2} = \frac{1}{27\cdot04} = \frac{1}{27}$ times as great.

Saturn:  $\frac{1}{9\cdot54^2} = \frac{1}{91}$ times as great.

Uranus:  $\frac{1}{19\cdot2^2} = \frac{1}{368\cdot6} = $ about $\frac{1}{370}$ times as great.

Neptune:  $\frac{1}{30^2} = \frac{1}{900}$ times as great.

These numbers are interesting, not merely because they give the apparent area of the sun as seen from the various planets, but also because they give a comparison between the amounts of light and heat from the sun which fall on similar areas of the planets. Any part of Mercury receives 6⅔ times as much light and heat as a similar spot on the earth, whereas remote Neptune receives but the $\frac{1}{900}$ part.

Before leaving these proportions and comparisons let us look for a comparison for the parallax that has to be measured when finding the distance of a star. The parallaxes that have to be measured are less than a second of arc. Let us think of a parallax of half a second, and of a halfpenny as a small measurable thing. At what distance must a halfpenny be placed, so that its angular width is a second? The equation is:

$$\sin \tfrac{1}{2}'' = \frac{\tfrac{1}{2} \text{ inch}}{x},$$

where $x$ is the distance at which the halfpenny must be placed.

$$x = \frac{\tfrac{1}{2} \text{ inch}}{\sin \tfrac{1}{2}''}.$$

Now $\qquad \sin \tfrac{1}{2}'' = \tfrac{1}{2}''$ in radians

$$= \cdot 000002424.$$

So $\qquad x = \dfrac{\tfrac{1}{2} \text{ inch}}{\cdot 000002424}$

$$= \text{about 200,000 inches}$$

$$= \text{about 3 miles.}$$

So if we imagine a halfpenny placed three miles away, and if we imagine ourselves pointing a telescope to its upper edge and then to its lower edge, and trying to measure the angle between, that is the kind of angle that has to be measured in finding the parallax of a near star.

The measurement for more distant stars is correspondingly more extraordinary. Arcturus has a parallax of ·008, which is less than $\frac{1}{80}$ of ·5; so we have to think of a halfpenny placed at a distance of more than 180 miles.

# CHAPTER 8

## *The Earth in a Balance*

OF ALL forlorn hopes that ever-hopeful man ever embarked upon, surely the most desperate was to attempt to weight the earth in a balance. Archimedes always had an eye on the gallery, and a saucy anecdote to popularise his discoveries. 'Scientist runs home naked to test theory.' 'Naked Scientist's Latest: Says he can shift the World.' But when Archimedes cried 'Give me a fulcrum, and I will move the world', he knew he was asking the impossible. The adventurer who set out to weigh the world in a small pair of scales meant to succeed, even expected to succeed, and strange to say did succeed.

What we aim to find is not strictly the weight of the earth, but its mass. We are all accustomed to judging masses by their weights. When two bits of stuff weigh the same we say they have the same mass, contain the same amount of matter. When we weigh tea or sugar it is the mass of material we are interested in; the weight is only a means of judging the mass. In an aeroplane doing a rapid turn the weight might suddenly bump up to six times as much. But even in the improbable event of tea being weighed on a spring balance and sold during such a manœuvre the purchaser would certainly refuse to pay six times the normal amount for his tea, because he would be getting no increase of mass.

Ordinary scales are a means of balancing masses. An increase or decrease of weight would not show because both sides of the balance would be affected alike; both might be increased by a tenth, for example, so that a normal 10 pounds on each side would be increased to 11 pounds on each side, without disturbing the balance.

It is a very different thing when we use a spring balance. A spring balance measures the actual weight of a thing; that is, it measures the force that draws out the spring. If the balance were taken up high above the earth it would indicate a small decrease in weight; and there are small but measurable differences in the weight of a body at one part of the earth's surface and its weight at other parts. Aeroplanes carry spring balances (accelerometers) to measure the enormous increases of weight that occur during rapid turns.

It is the unvarying mass of the earth we want to measure and not the variable weight. Indeed, it is difficult to attach any precise meaning to the expression 'the weight of the earth'.

It was Newton's investigation of the Law of Gravitation that made it possible to determine the mass of the earth.

According to Newton's Law the attraction between two bodies is directly proportional to the product of the masses of the two bodies. If we double one of the masses, we double the attraction; and that is what we should expect. If we double the other mass, we again double the attraction, so that it becomes four times as great. We express this by saying that the attraction varies as the product of the masses.

The attraction, according to Newton's Law, is also inversely proportional to the square of the distance between the centres of gravity of the two masses. If we double their distance apart, we reduce the attraction to a quarter; if we make the distance three times as great, we reduce the attraction to a ninth.

The Law of Gravitation is expressed most simply in symbols. We call the two masses whose attraction we are considering $M_1$ and $M_2$; the distance between the centres of gravity we call $d$. Then the attraction of each of the masses on the other is proportional to:

$$\frac{M_1 M_2}{d^2}.$$

Or, and this is another way of saying the same thing:

$$\text{attraction} = G \frac{M_1 M_2}{d^2}.$$

The constant $G$ in this equation is called the *constant of gravitation*. It is the same for the attraction between any two bodies, great or small. Actually it is the attraction between two bodies each of unit mass, whose centres of gravity are unit distance apart. The numerical value of the constant depends therefore on the particular units we are using.

There is one attraction that we are all familiar with. The attraction between a mass of one pound at the earth's surface and the earth itself, is one pound weight. One pound weight is a measure of the attraction between the earth and a mass of one pound at the earth's surface.

The pound weight is legally defined as the weight *in vacuo* of a platinum cylinder called the imperial standard pound. Actually this is a definition of a mass of one pound; a proper definition of an invariable pound weight would have to indicate the position of the platinum cylinder.

The first experiments to measure the mass of the earth were carried out by Henry Cavendish, the famous chemist. Let us look at the conditions in which he set to work. He had Newton's equation:

$$\text{1 pound weight} = G \frac{\text{1 pound mass} \times \text{mass of earth}}{(\text{radius of earth})^2}.$$

In this equation there are the standard quantities 1 pound weight and 1 pound mass; the radius of the earth was known. There remain the two unknowns, $G$ and the mass of the earth. If we can find one of these, then we have a simple equation to find the other. If we can measure $G$, we can find the mass of the earth.

Cavendish was not tied up to one particular example of Newton's equation; if he could find $G$ for any example, then he had found it for them all. He considered the equation:

$$\text{attraction} = G \frac{M_1 M_2}{d^2}.$$

He could weigh two bodies, and so find their masses. He could measure the distance $d$. All he needed was to find some way of measuring the attraction between the two bodies. If the weight of one of the bodies was balanced, the approach of the second body should produce a slight disturbance in its position; and the force producing this disturbance might turn out to be measurable. The attraction to be measured is obviously small; we are not normally aware of any gravitational attraction except that of the great mass of the earth. But there are ways of measuring even very small forces.

The original apparatus used by Cavendish was invented and constructed by the Rev. John Mitchell; Mitchell however did not live to carry out the experiments himself. The apparatus, as improved by Cavendish, consisted of a light wooden rod, about six feet long, with a ball of lead at each end. The rod was suspended by means of a thin wire at its mid-point. The apparatus was enclosed in a case to avoid the effects of air currents, and observations were made from a distance through a telescope. There was also a contrivance for bringing two large leaden balls close up to the balls at the ends of the rod; these large balls were outside the case; one was placed on each side, so that their attractions would cause rotation in the same direction.

The force required to move the rod, so as to twist the wire, could be found by giving the wire a small twist, and finding the time the end of the rod took to swing to and fro. The force required to produce any other small twist is proportional to the angle through which the rod, and therefore the wire also, is twisted.

When all was ready, the rod at rest, and the wire without twist, the exact position of the end of the rod was observed. The large leaden balls were brought into place, and the rod then moved through a small angle; this angle was measured. The large balls were then changed over to the opposite sides of the rod, and the angle was measured again.

In the actual experiments the small balls were 2 inches in diameter, and the large balls 12 inches, so that these were $6^3 = 216$ times as big and heavy as the small balls. The distances apart of the centres of gravity were rather less than 9 inches. It turned out that the attracting force was a very small fraction of a grain, which is itself the 7000th part of a pound weight.

Cavendish found for the value of $G$ 6.7, though the average of his twenty-nine measurements is rather higher. This is rather more than the result of extremely careful measurements by C. V. Boys in 1895 and Braun in 1896. These measurements give identical results which are now accepted as the value of $G$.

The value of $G$ found by Boys and Braun is, in c.g.s. units, $6.658 \times 10^{-8}$. That is the attraction in dynes between two masses of one gram at a distance apart of one centimetre.

In finding $G$ we multiply two masses and divide by the square of a distance. We have to keep this in mind when we change over to foot-pound-second units. A pound is 453.6 grams, so that the attraction of a pound is 453·6 times as great; this is the same for each pound, so we have to multiply by 453·6$^2$. A foot is 30·48 centimetres, so the attraction at a distance of one foot is reduced to $\dfrac{1}{30\cdot48^2}$. When we make these changes we have the attraction in dynes between two pound masses at a distance apart of one foot. To change to poundals we have to divide by 13,825, since 1 poundal = 13,825 dynes.

We now have:

$$G = 6\cdot658 \times 10^{-8} \text{ dyne}$$
$$= \frac{6\cdot658 \times 10^{-8} \times 453\cdot6^2}{30\cdot48^2 \times 13,825} \text{ poundal}$$
$$= 1\cdot067 \times 10^{-9} \text{ poundal.}$$

That is the attraction in poundals between two masses of one pound at a distance apart of one foot. If we want the attraction in pounds, we divide by 32·2.

$$G = 1\cdot067 \times 10^{-9} \text{ poundal}$$
$$= 1\cdot067 \times 10^{-9} \div 32\cdot2 \text{ pound}$$
$$= 3\cdot31 \times 10^{-11} \text{ pound.}$$

Let us go back to the equation:

$$\text{attraction} = G\,\frac{M_1 M_2}{d^2};$$

or: $\quad$ weight of 1 pound $= G\,\dfrac{\text{mass of 1 pound} \times \text{mass of earth}}{(\text{radius of earth})^2}$.

We know that 1 pound weight $= g \times$ mass of 1 pound, or 32·2 times the mass of 1 pound. We can take the radius of the earth as 3960 miles $= 3960 \times 5280$ feet.

The equation becomes:

$$32\cdot2 \times 1 \text{ pound mass} = G\,\frac{\text{mass of 1 pound} \times \text{mass of earth}}{3960^2 \times 5280^2}.$$

The pound mass cancels out, and we find:

$$\text{mass of earth} = \frac{32\cdot2 \times 3960^2 \times 5280^2}{G} \text{ pounds}$$
$$= \frac{32\cdot2 \times 3960^2 \times 5280^2}{1\cdot067 \times 10^{-9} \times 2240} \text{ tons}$$
$$= 5\cdot89 \times 10^{21} \text{ tons,}$$

or nearly 6000 trillion tons.

That is the mass that Cavendish set out to find.

If we use the c.g.s. value of $G$, we have:

$$1 \text{ gram weight} = G \frac{1 \text{ gram mass} \times \text{mass of earth}}{(\text{radius of earth})^2}.$$

1 gram weight $= g$ times 1 gram mass, and $g$ in c.g.s. units is 981 centimetres per second per second.

We also want the earth's radius in centimetres.

$$3960 \text{ miles} = 3960 \times 5280 \times 12 \text{ inches}$$
$$= 3960 \times 5280 \times 12 \times 2 \cdot 54 \text{ centimetres}$$
$$= 6 \cdot 372 \times 10^8 \text{ centimetres}.$$

The equation becomes:

$$981 \times 1 \text{ gram mass} = G \frac{1 \text{ gram mass} \times \text{mass of earth}}{(\text{radius of earth})^2}$$

$$= \frac{6 \cdot 658 \times 10^{-8} \times 1 \text{ gram mass} \times \text{mass of earth}}{6 \cdot 372^2 \times 10^{16}}.$$

1 gram mass cancels out, and we have:

$$\text{mass of earth} = \frac{6 \cdot 372^2 \times 10^{16} \times 10^8 \times 981}{6 \cdot 658} \text{ grams}$$

$$= 5 \cdot 98 \times 10^{27} \text{ grams}.$$

We can readily change this into tons.

$$1 \text{ gram} = \cdot 0022046 \text{ pound},$$

$$5 \cdot 98 \times 10^{27} \text{ grams} = \frac{5 \cdot 98 \times \cdot 0022046 \times 10^{27}}{2240} \text{ tons}$$

$$= 5 \cdot 89 \times 10^{21} \text{ tons}.$$

The book of *Constants* gives the values as

$$5 \cdot 98 \times 10^{27} \text{ grams} = 5 \cdot 87 \times 10^{21} \text{ tons}.$$

As may be seen, these two values are not quite equal.

We can use the measured and calculated mass of the earth to find its average density. We will find it first in tons per cubic yard. We have merely to divide the mass by the number of cubic yards in the earth. 'Per' can be read as an instruction to divide.

The volume of the earth is:

$$\tfrac{4}{3}\pi r^3 = \tfrac{4}{3} \times 3 \cdot 142 \times (3960 \times 1760)^3 \text{ cubic yards.}$$

The average density is:

$$\frac{5 \cdot 89 \times 10^{21}}{\frac{4}{3} \times 3 \cdot 142 \times (3960 \times 1760)^3} \text{ tons per cubic yard}$$

$$= 4 \cdot 154 \text{ tons per cubic yard.}$$

$$\left( \cdot 154 = \frac{154}{1000} = \frac{1}{6 \cdot 5} = \frac{2}{13} \cdot \right)$$

So that the average density of the earth is about $4\frac{2}{13}$ tons per cubic yard. The density of broken coal is about 1 ton per cubic yard, so that the density of the earth is very high.

We will now find the density in pounds per cubic foot.

$4 \cdot 154$ tons per cubic yard $= 4 \cdot 154 \times 2240$ pounds per cubic yard

$$= \frac{4 \cdot 154 \times 2240}{27} \text{ pounds per cubic foot}$$

$$= 344 \cdot 3 \text{ pounds per cubic foot.}$$

The density of water is $62 \cdot 3$ pounds per cubic foot, so that the average density of the earth is:

$$\frac{344 \cdot 3}{62 \cdot 3} = 5 \cdot 527 \text{ times the density of water,}$$

or a little more than $5\frac{1}{2}$ times as great.

Cavendish actually found $5 \cdot 48$ for the density of the earth, but recent measurements, using the same 'torsion balance' method, place the density at $5 \cdot 527$ or even a little higher.

Slightly lower results are given by the common balance method. The principle of this method is quite simple. A heavy mass of lead is exactly balanced on a common balance. Another mass of lead is then held a short distance above it. The attraction between the two masses helps to support the lower mass, and a small weight must be removed from the other pan to restore the balance. This weight measures the attraction between the two masses, and so enables the calculation to proceed as before.

The density found for the earth as a whole is just about double the density of rocks near the surface. Granite has a density between $2\frac{1}{2}$ and 3 times that of water; basalt is about 3; sandstone is about $2\frac{1}{4}$. On the other hand, the densities of most of the common metals are about half as big again as the density of the earth.

Just as we need an earthly measurement of length to cast a line over the solar system, so the measurement of the mass of the earth enables us to find the masses of the sun, moon and planets; though we can find these masses, using the earth's mass as unit, without actually measuring the mass of the earth.

We have seen that the distances of the planets from the sun can be measured and calculated. The time of revolution of a planet about the sun, or of a satellite about a planet, can be measured with great accuracy.

Now if $d$ be the mean distance of a planet from the sun (its semi-major axis), and $t$ its periodic time (the length of its sidereal year), then we have an equation between these two. It is:

$$\frac{d^3}{t^2} = c\,(M+m).$$

In this equation $c$ is a constant. $M$ and $m$ are the masses of the sun and a planet which we are considering. The masses of the planets are so small compared with that of the sun that we can usually ignore $m$ and say that $\frac{d^3}{t^2} = cM$, so that it is a constant, $M$ being of course the same for all the planets. If we were dealing with another system, say the moons of Jupiter, we should have a different constant because it would include the mass of Jupiter instead of the mass of the sun.

For the earth in particular we have:

$$\frac{d^3}{t^2} = \frac{92 \cdot 9^3}{(365\frac{1}{4})^2} \quad \text{(cube of distance in millions of miles)} \atop \text{(square of periodic time in days)}.$$

This quantity is equal to $c$ times the mass of the sun, if we ignore the mass of the earth.

$$c \times \text{mass of sun} = \frac{92 \cdot 9^3}{(365\frac{1}{4})^2}.$$

For the moon revolving round the earth we have:

$$\frac{d^3}{t^2} = \frac{\cdot 2389^3}{(27\frac{1}{3})^2} \quad \text{(cube of distance from earth)} \atop \text{(square of periodic time in days)}.$$

This is equal to $c$ times the mass of the earth, if we ignore the mass of the moon.

$$c \times \text{mass of earth} = \frac{\cdot 2389^3}{(27\frac{1}{3})^2}.$$

We can now divide the first equation by the second, and so get rid of $c$.

$$\frac{c \times \text{mass of sun}}{c \times \text{mass of earth}} = \frac{92 \cdot 9^3}{(365\frac{1}{4})^2} \div \frac{\cdot 2389^3}{(27\frac{1}{3})^2}$$

$$= \frac{92 \cdot 9^3 \times (27\frac{1}{3})^2}{(365\frac{1}{4})^2 \times \cdot 2389^3}$$

$$= 324,300.$$

So far as this calculation goes it appears that the sun has a mass something like 324,300 times as great as that of the earth, so that we are quite justified in ignoring the mass of the earth as part of the total mass: earth + sun. We should however include the mass of the moon with that of the earth, so that 324,300 is actually to be multiplied by the combined mass of earth and moon. We know from other sources that the moon has rather less than $\frac{1}{80}$ of the mass of the earth, so that we should get a more accurate result by adding an eightieth of the first approximation.

$$324,300 + \tfrac{1}{80} \times 324,300 = 324,300 + 4050$$
$$= 328,350.$$

That is, the sun has a mass equal to about 330,000 times that of the earth.

If we happen to want the mass of the sun in tons, we can readily find it from the measured mass of the earth. It is:

$$5 \cdot 89 \times 10^{21} \times 330,000 \text{ tons}$$
$$= 19 \cdot 44 \times 10^{26} \text{ tons, or } 1 \cdot 944 \times 10^{27} \text{ tons,}$$

or about 2000 quadrillion tons. Set out in full it is:

$$1,944,000,000,000,000,000,000,000,000 \text{ tons.}$$

The first way of writing the mass is distinctly to be preferred to the last.

We have seen that the sun has a volume equal to about 1,300,000 times that of the earth, whereas its mass is only 330,000 times as great; so the sun is evidently much less dense than the earth. Its density is:

$$\frac{330,000}{1,300,000} = \frac{33}{130} = \text{about } \tfrac{1}{4} \text{ of that of the earth.}$$

The density of the sun is therefore a little less than half as much again as the density of water. The sun is therefore just about as dense as an ordinary piece of coal.

It is something of a surprise to realise that the moon is always falling toward the earth without getting any nearer to it. The apparently contradictory result is of course due to the forward movement of the moon.

In the diagram $E$ is the earth, and $M$ is the moon moving in its nearly circular orbit. If the moon and the earth were to cease to attract each other, we know that the moon would move off along the tangent, $MA$, to its orbit. The pull of the earth causes the moon to fall toward the earth, and so to remain on the orbit. The fall is represented by $AB$. Actually the moon would be farther round in its orbit than $B$, as may be seen by wrapping $MA$ round the orbit; but for short distances the difference is insignificant, and we only want short distances.

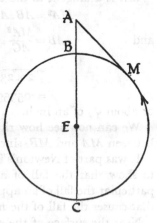

We can readily find the speed of the moon in its orbit. We shall not be far out if we treat the moon's orbit round the earth as a circle of radius 238,900 miles. The circumference is:

$$2\pi r = 2 \times 3 \cdot 142 \times 238,900 \text{ miles.}$$

The moon traverses its orbit in about $27\frac{1}{3}$ days, so the average speed is:

$$\frac{2 \times 3 \cdot 142 \times 238,900}{27\frac{1}{3}} \text{ miles per day}$$

$$= \frac{2 \times 3 \cdot 142 \times 238,900 \times 5280 \times 12}{27\frac{1}{3} \times 24 \times 3600} \text{ inches per second}$$

$$= 40,280 \text{ inches per second.}$$

That is, 3357 feet per second, or $\frac{15}{22} \times 3357 = 2289$ miles per hour. But we actually want to use the speed in inches per second.

For such a short movement in the orbit as roughly $\frac{2}{3}$ of a mile we can take the tangent $MA$ as being equal to the distance the moon moves. That is, $MA = MB = 40,280$ inches.

Also $AC = BC$ (almost exactly) = the diameter of the moon's orbit

$$= 2 \times 238{,}900 \text{ miles}$$
$$= 2 \times 238{,}900 \times 5280 \times 12 \text{ inches,}$$

since we are working in inches.

$MA$ is a tangent to the orbit, so:

$$MA^2 = AB.AC$$

and

$$AB = \frac{MA^2}{AC}$$

$$= \frac{40{,}280^2}{2 \times 238{,}900 \times 5280 \times 12} \text{ inch}$$

$$= \cdot05359 \text{ inch,}$$

or about $\frac{1}{18}$ of an inch.

We can now see how right we were in ignoring the difference between $MA$ and $MB$, since the fall $AB$ is only $\frac{1}{18}$ inch in $\frac{2}{3}$ mile.

It was part of Newton's investigation into universal gravitation to show that the fall of an object near the earth's surface (in particular the fall of an apple from a tree) is due to the same force that causes the fall of the moon toward the earth.

Near the surface of the earth an object falls $16 \cdot 1$ feet (= $193 \cdot 2$ inches) in the first second. The moon is:

$\frac{238{,}900}{3960}$ times as far from the earth's centre as is an object at the earth's surface       $= 60 \cdot 3$ times as far.

If gravity fades away according to the inverse square law, as Newton imagined, then it should be $\frac{1}{60 \cdot 3^2}$ times as great in its effect on the moon, as on an object near the earth's surface. If this law is true, the fall of the moon toward the earth in one second should be:

$$\frac{193 \cdot 2}{60 \cdot 3^2} \text{ inch} = \cdot0531 \text{ inch.}$$

Allowing for errors of measurement, and for several approximations, this result agrees extremely well with the measured and calculated distance of $\cdot05358$ inch. The inverse square law can be taken as proved for the effects of gravity between earth and moon. The generalisation that the law is universal can only be accepted if we find that it is true for every case that is investigated. Discrepancies might lead to a modification of the law. So far as large bodies like the planets are concerned the law as it stands is satisfactory, or almost entirely so. The law breaks down altogether when attempts are made to apply it to small fast-moving particles. The reason is that effects which are minute in large bodies moving with comparative slowness become supremely important in very fast particles.

Just as the moon falls toward the earth, so the earth falls toward the sun, and we can apply the same method to investigate this fall. We begin by finding the speed of the earth in its orbit. The circumference is about:

$$2\pi r = 2 \times 3 \cdot 142 \times 92 \cdot 9 \times 10^6 \text{ miles.}$$

The average speed in the orbit is:

$$\frac{2 \times 3 \cdot 142 \times 92 \cdot 9 \times 10^6}{365\frac{1}{4}} \text{ miles per day}$$

$$= \frac{2 \times 3 \cdot 142 \times 92 \cdot 9 \times 10^6 \times 5280 \times 12}{365\frac{1}{4} \times 24 \times 3600} \text{ inches per second}$$

$$= 1,171,000 \text{ inches per second.}$$

That is, about 98,000 feet per second, or about 67,000 miles per hour. As before, we want to use the speed as inches per second.

In the diagram $E$ is the earth moving in a nearly circular orbit about the sun. $AE$ is to represent the distance moved by the earth in a second, and we have the same argument as before for ignoring the fact that $AE$ and $BE$ are not strictly equal. Also we take $AC = BC =$ the diameter of the earth's orbit $= 2 \times 92 \cdot 9 \times 10^6$ miles $= 2 \times 92 \cdot 9 \times 10^6 \times 5280 \times 12$ inches.

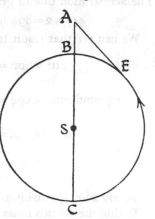

We want to find $AB$, the fall toward the earth in a second. We have:

$$AE^2 = AB \cdot AC,$$

$$AB = \frac{AE^2}{AC}$$

$$= \frac{1,171,000^2}{2 \times 92 \cdot 9 \times 10^6 \times 5280 \times 12} \text{ inch per second}$$

$$= \cdot 1166 \text{ inch per second.}$$

We have this minute fall, of not much more than a ninth of an inch in 98,000 feet, or $18\frac{1}{2}$ miles, to change a straight path into the curve of the earth's orbit.

The radius of the sun is 433,000 miles; the distance of the earth from the sun is 92.9 million miles. So that the earth is:

$$\frac{92,900,000}{433,000} = 214\cdot55$$

times as far from the centre of the sun as is a point on the sun's surface. The attraction of the sun is therefore $214\cdot55^2$ times as great at its own surface, as at the distance of the earth.

$$214\cdot55^2 = 46,000.$$

Hence an object at the sun's surface would fall 46,000 times as far in the first second as the earth falls in a second. That is:

$$\cdot1166 \times 46,000 = 5364 \text{ inches}$$
$$= 447 \text{ feet in the first second.}$$

Gravity on the sun's surface is:

$$\frac{447}{16\cdot1} = 27\cdot77 \text{ times as great as at the earth's surface.}$$

The acceleration due to gravity on the sun is:

$$447 \times 2 = 894 \text{ feet per second per second.}$$

We can use that result to find the mass of the sun:

$$\text{attraction} = G\,\frac{M_1 M_2}{d^2},$$

$$1 \text{ pound mass} \times 894 = \frac{G \times 1 \text{ pound mass} \times \text{mass of sun}}{(\text{radius of sun})^2},$$

$$\text{mass of sun} = \frac{894 \times (432,900 \times 5280)^2}{1\cdot067 \times 10^{-9} \times 2240} \text{ tons}$$
$$= 1\cdot96 \times 10^{27} \text{ tons}$$
$$= 1960 \text{ quadrillion tons,}$$

or nearly 2000 quadrillion tons.

Taking the earth's mass as unity we have:

$$\frac{1\cdot96 \times 10^{27}}{5\cdot89 \times 10^{21}} = 333,000.$$

So that the sun appears to have 333,000 times the mass of the earth.

The mass of a planet which has a satellite can be found in the same way as that used for the earth and the sun. One of Jupiter's satellites, for example, revolves about the planet in $42\frac{1}{8}$ hours at a distance of 265,300 miles. The length of the orbit is about:

$$2\pi r = 2 \times 3{\cdot}142 \times 265{,}300 \text{ miles.}$$

The speed in the orbit is:

$$\frac{2 \times 3{\cdot}142 \times 265{,}300}{42\frac{1}{8}} \text{ miles per hour}$$

$$= \frac{2 \times 3{\cdot}142 \times 265{,}300 \times 5280 \times 12}{42\frac{1}{8} \times 3600} \text{ inches per second}$$

$$= 693{,}000 \text{ inches per second.}$$

Following the same process as before, we find the fall toward Jupiter to be:

$$\frac{693{,}000^2}{\text{diameter of orbit in inches}} \text{ inches per second}$$

$$= \frac{693{,}000^2}{2 \times 265{,}300 \times 5280 \times 12} \text{ inches per second}$$

$$= 14{\cdot}29 \text{ inches per second.}$$

That result is much greater than for the earth and moon because the speed of revolution is much greater.

The radius of Jupiter is 43,850 miles, so the satellite is:

$$\frac{265{,}300}{43{,}850} = 6 \text{ times the radius of Jupiter distant from the centre of}$$
the planet.

Hence gravity at the surface of Jupiter is $6^2 = 36$ times as great as at the distance of the satellite. That is, an object near the surface falls in the first second:

$$14{\cdot}29 \times 36 \text{ inches}$$
$$= 42{\cdot}9 \text{ feet.}$$

Thus gravity at the surface of Jupiter is:

$$\frac{42{\cdot}9}{16{\cdot}1} = \text{about } 2{\cdot}6 \text{ times what it is on the earth.}$$

A more accurate result is $2{\cdot}57$.

Now Jupiter's radius is $11{\cdot}06$ times the radius of the earth. If the masses of the earth and Jupiter were equal, gravity on Jupiter would be $\frac{1}{11{\cdot}06^2}$ of what it is on the earth. The mass of

Jupiter must be great enough to cancel this factor, and also to increase gravity 2·57 times. Hence the mass of Jupiter is:

$$11·06^2 \times 2·57 = 314·3 \text{ times the mass of the earth.}$$

Jupiter is not a dense planet. Its radius is 11·06 times that of the earth, so that its volume is:

$$11·06^3 = 1352 \text{ times as great.}$$

Its density therefore can only be:

$$\frac{314·3}{1352} = \text{about } \tfrac{1}{4} \text{ as great.}$$

When we have found the mass and radius of a planet or a satellite we have the information necessary to find the value of gravity at its surface. The radius of Mercury is given as 1387 miles, and its mass as ·04 of the mass of the earth, or

$$·04 \times 5·89 \times 10^{21} \text{ tons}$$
$$= 2·356 \times 10^{20} \text{ tons.}$$

Weight of 1 pound mass on Mercury

$$= G \frac{1 \text{ pound mass} \times \text{mass of Mercury}}{(\text{radius of Mercury})^2}.$$

(We want the mass in pounds and the radius in feet. Then the weight will be in poundals.)

$$= \frac{1·067 \times 10^{-9} \times 2·356 \times 10^{20} \times 2240}{(1387 \times 5280)^2}$$
$$= 10·5 \text{ poundals.}$$

Hence gravity on Mercury is:

$$\frac{10·5}{32·2} = ·32 \text{ times what it is on the earth.}$$

We can get the comparison with gravity on the earth more quickly. The mass of Mercury (earth = 1) is ·04; the radius of Mercury (earth = 1) is ·35.

$$\frac{\text{gravity on Mercury}}{\text{gravity on earth}} = \frac{·04}{·35^2} = ·32.$$

When we say that gravity on the sun is about 27·8 times what it is on the earth we include two important effects; similar objects would weigh 27·8 times as much at the surface of the sun as on the earth, and they would fall with 27·8 times the speed; that is, 448 feet in the first second of fall from rest, with an acceleration of 896 feet per second per second. A fall of 2 feet on the sun would produce the same speed of descent as a fall of 55 feet on the earth. Whereas on Mercury, where gravity is only about $\frac{3}{10}$ of what it is on the earth, a body would weigh $\frac{3}{10}$ of what it does on the earth, it would fall with $\frac{3}{10}$ of the acceleration, and a fall of 30 feet on Mercury would do no more harm than a fall of $\frac{3}{10} \times 30 = 9$ feet on the earth.

I have tried to keep clear the distinction between mass (the amount of matter in a body) and its weight (the force with which the earth pulls it). Weight is always relative to something. In the examples with which we are most familiar it is relative to the mass of the earth and distance from the centre of the earth. An aeroplane flying at a height of 5 miles would weigh less than on the earth. Weight varies inversely as the square of the distance, so it would be reduced to:

$$\frac{3960^2}{3965^2} = \cdot99748 \text{ times the weight at sea-level.}$$

The difference is 252 parts in 100,000. In a weight of 10 tons that would be a reduction of:

$$\cdot0252 \text{ ton} = \cdot0252 \times 2240 \text{ pounds}$$
$$= 56\cdot4 \text{ pounds or over 4 stones.}$$

At a height of 10 miles the reduction would be:

$$\frac{3960^2}{3970^2} = \cdot99497.$$

That is 503 parts in 100,000, and in a weight of 10 tons it is 112·7 pounds, or just over a hundredweight.

At a height of 100 miles above the earth's surface the reduction in weight would be:

$$\frac{3960^2}{4060^2} = \cdot9513.$$

That is 487 parts in 10,000, or 1091 pounds in 10 tons = nearly 10 hundredweight in 10 tons.

At 4000 miles' distance, twice the length of the earth's radius from the earth's centre, the weight would be reduced to a quarter. At the distance of the moon, 60 times the earth's radius, weight would be reduced to $\frac{1}{60^2} = \frac{1}{3600}$. That is, weight relative to the earth.

Somewhere between the earth and the moon is a point where the earth's pull is exactly balanced by the pull of the moon. Objects in this region have no weight either toward the earth or toward the moon. It happens to be very easy to fix this point. The moon has $\frac{1}{81}$ of the mass of the earth; so the distance from the moon is $\frac{1}{\sqrt{81}} = \frac{1}{9}$ of the distance from the earth. The distance from the earth is $\frac{9}{10} \times 239,000 = 215,100$ miles; and the distance from the moon is $\frac{1}{10} \times 239,000 = 23,900$ miles.

Jules Verne and H. G. Wells have both tried to imagine what would happen to weightless men. Jules Verne treats the matter rather crudely by letting the change come suddenly. People and objects are suspended in space, and then settle down again toward the moon side. The change would however come slowly. A 10-stone man, at a distance of 40,000 miles from the centre of the earth, would weigh $\frac{1}{10^2} = \frac{1}{100}$ of 10 stone, or about $1\frac{1}{2}$ pounds. In the next 175,000 miles of the journey his weight would decrease more and more slowly from $1\frac{1}{2}$ pounds down to zero. Then it would increase again up to $\frac{1}{6}$ of 10 stone at the moon's surface.

H. G. Wells lets his heroes feel the tug of the moon when they are only 800 miles from the earth, or say 238,000 miles from the moon. (The earth's attraction is supposed to be screened off.) That is:

$$\frac{238,000}{1080} = \text{about 220 times the moon's radius away.}$$

So that gravity would be $\frac{1}{220^2} = \frac{1}{48,400}$ of what it is at the moon's surface, or $\frac{1}{48,400 \times 6}$ of what it is at the earth's surface:

$$= \text{about } \frac{1}{290,000} = \cdot 00000345.$$

The weight of a 12-stone man would be:

$12 \times 14 \times 16 \times \cdot 00000345$ ounce $= $ about $\frac{1}{100}$ ounce.

The acceleration due to gravity would be:

$32 \cdot 2 \times \cdot 00000345$ foot per second per second

$= $ about $\cdot 0001$ foot per second per second,

or a little more than a ten-thousandth of an inch per second per second.

They must have been very acute observers to feel so minute a weight, and to perceive the motion as 'creeping slowly down the glass'. The time taken to fall a foot would be:

$$s = \tfrac{1}{2}gt^2,$$
$$1 = \tfrac{1}{2} \times \cdot 0001 t^2,$$
$$t = 140 \text{ seconds, or over two minutes.}$$

But I am more troubled about the attraction between the two men. Taking them to be 5 feet apart, and each to weigh 12 stones, the attraction in poundals would be:

$$G\,\frac{M_1 M_2}{d^2} = \frac{1 \cdot 067 \times 10^{-9} \times 168 \times 168}{5^2}$$

$$= \cdot 0000012 \text{ poundal}$$
or
$$\cdot 00000004 \text{ pound.}$$

The attraction is about 4 hundred-millionths of a pound.

Gravity on the earth on one of these two masses of 12 stones would be  $12 \times 14 = 168$  pounds. So gravity is reduced to $\cdot 00000004 \div 168 = $ about $\cdot 00000000024$ of what it is on the earth's surface, or $2 \cdot 4 \times 10^{-10}$. The acceleration would be:

$$32 \cdot 2 \times 2 \cdot 4 \times 10^{-10} \text{ feet per second per second}$$
$$= 7 \cdot 7 \times 10^{-9} \text{ foot per second per second.}$$

The bodies would fall the $2\tfrac{1}{2}$ feet toward the common centre in:

$$s = \tfrac{1}{2}gt^2,$$
$$2\tfrac{1}{2} = \tfrac{1}{2} \times 7 \cdot 7 \times 10^{-9}t^2,$$
$$t^2 = \frac{5}{7 \cdot 7} \times 10^9,$$
$$t = \sqrt{\frac{50}{7 \cdot 7}} \times 10^4$$
$$= \text{about } 25,500 \text{ seconds.}$$

That is, more than 7 hours. I can hardly think a fall of two feet in seven hours would be observable, especially in the much slower initial stages.

In one of Mr Shaw's *Prefaces* I came across the following curious statement: 'The modern man who believes that the earth is round is grossly credulous. Flat Earth men drive him to fury by confuting him with the greatest ease when he tries to argue about it. Confront him with a theory that the earth is cylindrical, or annular, or hour-glass shaped, and he is lost. The thing he believes may be true, but that is not why he believes it: he believes it because in some mysterious way it appeals to his imagination. If you ask him why he believes the sun is ninety-odd million miles off, either he will confess that he doesn't know, or he will say that Newton proved it. But he has not read the treatise in which Newton proved it, and does not even know that it was written in Latin. If you press an Ulster Protestant as to why he regards Newton as an infallible authority...no reasonable reply is possible.' There is a lot more in the middle of the last sentence, but that seems to be enough.

The statement is curious because it seems to be another of those statements which contain every error it is possible to make on the subject with which it deals. Mr Shaw must have a strange circle of acquaintances if they are driven to fury by Flat Earthists; there can hardly be a single seafarer amongst them, or even one who has cast an intelligent eye on the circle of the horizon, or intelligently observed an eclipse of the moon. If Mr Shaw has confronted ordinary people with his ideas of a cylindrical, or annular or hour-glass shaped earth, they may have appeared lost because they were wondering whether they should humour him. Is it necessary to say that Newton did not prove that the sun is ninety-odd million miles away? One does not prove distances, one measures them. And in any case the measurement of the sun's parallax was not made by Newton; 'further to this' it was the moon's distance that Newton was specially interested in. I suppose the reference to Newton's treatise is meant for the *Principia*; every schoolboy learns that it was written in Latin, and, for all his air of superiority, how much more does Mr Shaw know about it?

It is in his idea that Newton is regarded as an infallible authority that Mr Shaw errs most grievously. The idea is Mr Shaw's own. He tries to place his cuckoo's egg in the nest of an 'Ulster Protestant'; but I doubt whether anyone would be willing to act as foster parent.

Probably no scientist or mathematician has been regarded as infallible; at any rate, not since Aristotle's centuries-long reputation began to fade. Newton was certainly not so regarded; he

was far too great a man to be infallible. He submitted to the
scientific world a reasoned argument to prove the probability of
his gravitational equation. It was the argument that was
considered, and criticised, and accepted as valid. Not Newton's
personality.

I am not trying to shift the idea of infallibility from the man
to the argument. Newton's idea of gravitation has endured
because it accords with facts. But if new facts come to light that
do not accord with the theory, then the theory has to be modified.
Newton's theory has stood the test of time remarkably well; for
most purposes it is all that is required, and it has the great virtue
of simplicity. Nevertheless there have been discrepancies, and
modifications of the attraction equation were proposed even
before Einstein; one of these was to replace the square of the
distance by a power a very little less than 2.

# CHAPTER 9

## *The Prodigal Sun*

WE WERE all taught as children that matter is indestructible. It might be changed in many ways. We might have physical changes by reducing matter to powder, by crystallising it, freezing it, melting it, condensing it, evaporating it. We might have chemical changes by causing components to unite in various ways. We might see matter disappear in solution, or vanish in flames. Throughout all the changes, physical and chemical, the whole amount of matter concerned in the changes under observation remained the same. This was established experimentally in thousands of cases by very careful weighing. And so we arrived at the idea of the *indestructibility of matter*, or the *conservation of mass*. The principle was universally accepted. It formed, and still forms, the major premiss of quantitative chemistry. Indeed the principle went further: it was tacitly assumed that the elements are immutable, and that the total amount of each element does not vary.

In mechanics we had the corresponding idea of the *conservation of energy*. Energy was conceived to be just as indestructible as matter. It might be changed in various ways. The kinetic energy of falling water may be changed to the energy of a rotating wheel, this again to the rotation of coils in a dynamo, then to electric energy, the rotation of an electric motor, and so on. Throughout all these changes the total amount of energy is unchanged.

The principle of the conservation of energy was not so readily perceived as that of the conservation of mass. First perceptions seemed to show that there is no such principle. We commonly speak of energy being 'used up' in overcoming resistances; at every stage in the process of transforming energy there appears to be a loss of energy. It is a commonplace of mechanics that we never get out of a machine more than we put into it; we actually get less, because there is always some energy used in overcoming friction inside the machine. The advantage of a machine is not that it creates energy, but that it changes the energy to a more usable form.

The principle of the conservation of energy could not be accepted until the apparent loss of energy was traced to the change of kinetic energy into heat, the heat that is produced when two surfaces rub together, or when an electric current flows through a wire. Heat itself is a form of motion (as Tyndall announced in the title of his book *Heat a Mode of Motion*). The heat of a body consists in oscillations of the molecules composing it; the higher the temperature the more rapid are the oscillations. It should not be a matter for surprise therefore that kinetic energy can be transformed to heat. Joule succeeded in measuring the 'mechanical equivalent of heat', that is, the amount of mechanical energy equivalent to a unit of heat. The principle of his method was to perform a measurable quantity of mechanical work, to find the effect of this in raising the temperature of a quantity of water, and so to find the number of units of heat into which the mechanical energy had been transformed. Joule showed that it takes about 778 foot-pounds of energy to raise one pound of water one degree Fahrenheit. To bring a pint of ice-cold water (which weighs $1\frac{1}{4}$ pounds) to boiling-point would require:

$$778 \times 1\frac{1}{4} \times 180 = 175,000 \text{ foot-pounds}$$

One horse-power $= 33,000$ foot-pounds per minute.

$$\frac{175,000}{33,000} = \text{about } 5\frac{1}{4}.$$

So one horse-power, completely transformed into heat, would boil a pint of ice-cold water in $5\frac{1}{4}$ minutes.

The two principles of the conservation of mass and the conservation of energy have now been replaced by the single principle of the conservation of the combined total of mass and energy. We have become familiar with the idea that mass can be changed into energy, and that there is an exact relation between the amount of matter changed and the amount of energy into which it is changed. This idea is fundamental to modern physics.

Ordinary chemical reactions are concerned with the electrons which form a very small part of the mass of an atom. The mass is almost entirely concentrated in the very minute nucleus, and it is in the nucleus that there is a vast reservoir of energy. A change in the nucleus of an atom involves its transmutation into an atom of another element. It is in such nuclear reactions that energy is released. Nuclear reactions have been brought about, but only on a minute scale and with the expenditure of a disproportionate amount of energy to bring about the transformations.

The modern principle has been expressed in an equation:

$$e = mc^2.$$

In this equation $e$ is the energy produced by the transformation of a certain amount of matter, $m$ is the mass of the matter transformed and $c$ is the speed of light. All these quantities have to be expressed in the appropriate units.

The form of the equation, with the square of the exceedingly high speed of light, should prepare us for the fact that the energy released by converting even a small mass into energy is very great. Let us look at the energy that could be obtained by the transformation of a single pound mass.

If we use feet and pounds as the units, the energy will be in foot-poundals. We have to change the speed of light to feet per second:

186,000 miles per second = 186,000 × 5280 feet per second.

The amount of energy from a single pound is:

$$e = mc^2$$
$$= 1 \text{ pound mass} \times (186{,}000 \times 5280)^2$$
$$= 9 \cdot 64 \times 10^{17} \text{ foot-poundals,}$$

or ·964 trillion foot-poundals, which is not far short of a trillion foot-poundals.

This is evidently a vast amount of energy. We can get a better picture of it by turning it into horse-power.

1 horse-power
= 33,000 foot-pounds per minute
= 32 × 33,000 foot-poundals per minute
= 32 × 33,000 × 60 × 24 × 365¼ foot-poundals per year.

We want to know how many horse-power for a year could be

obtained from one pound mass. We have only to work out the division:

$$\frac{\text{number of foot-poundals from a pound}}{\text{number of foot-poundals per year in 1 horse-power}}$$

$$= \frac{\cdot 964 \times 10^{18}}{32 \times 33,000 \times 60 \times 24 \times 365\frac{1}{4}} \text{ horse-power}$$

$$= 1\cdot 736 \times 10^6 \text{ horse-power}$$

$$= \text{about } 1\tfrac{3}{4} \text{ million horse-power.}$$

That is to say, the conversion of a pound of mass into energy would provide the equivalent of $1\frac{3}{4}$ million horses working continuously at full pressure, night and day, for a year.

The total horse-power developed by the Niagara Falls is about $3\frac{1}{2}$ million. Now $3\frac{1}{2} \div 1\frac{3}{4} = 2$. So that the transformation of 2 pounds of mass into energy every year would supply the same amount of energy as the Niagara Falls. It is extraordinary to think that the energy of that profligate waste of water, thundering down for a whole year, should be latent in two pounds of apparently inert matter.

There is no means that we know of for converting a comparatively large mass like a pound into energy: that is to say, there is no terrestrial means. If there were a simple and controllable method, then all our energy problems would be solved. A large steam turbine supplies perhaps 40,000 horse-power from a great expenditure of fuel. If we could release the nuclear energy, we should need to transform:

$$\frac{40,000}{1\frac{3}{4} \times 10^6} \text{ of a pound each year to obtain this horse-power continuously}$$

$$= \frac{4}{175} \text{ pound}$$

$$= \frac{64}{175} \text{ ounce} = \frac{1}{2\cdot 7} \text{ ounce,}$$

or rather more than a third of an ounce each year.

The source of the sun's vast supplies of energy, which it expends so lavishly in light and heat and other forms of radiation, has long been a matter for speculation. The old idea that heat and light are supplied merely by the cooling down of a very hot body, turned out to be untenable. Lord Kelvin showed that if that were indeed the sole source of light and heat, then the sun could not be much more than 20 million years old. And such a period was quite inadequate to account for the great geological and biological changes that have taken place on the earth.

We need a far greater source of energy to account for the observed facts. The equation connecting mass and energy suggests what this source is. We have seen the vast amount of energy that is released by the transformation of a mere pound of matter. On a very minute scale the transformation has been accomplished in laboratories; it occurs spontaneously in the disintegration of radium and other radioactive elements. The enormously high temperatures in the interior of the sun and of the stars effect the transformation on a large scale.

We have a means of measuring the solar radiation, that is the whole amount of heat and light radiated by the sun; so that we have a measure of the vast size of the problem. The original experiments were carried out with an instrument called a pyro-heliometer (sun-heat measurer). It consisted of a square of dull blackened copper on which sunlight was allowed to fall. Behind this blackened square water circulated slowly and evenly. The dull black copper surface absorbed practically all the light and heat that fell on it, and this radiation passed on to the water. The water slowly rose in temperature, and after a time it issued at a steady temperature. The change in temperature of the circulating water, between its entrance and its exit, gave a measure of the radiation absorbed. Allowance had to be made for the fact that the instrument did not receive the full solar radiation; part of it was absorbed by the atmosphere. Allowance for absorption by the atmosphere was made by taking observations at different levels, some of them as high up as possible. As a final result it was found that the amount of heat that would be absorbed by a square centimetre of the earth's surface, if there were no atmosphere, is 2 calories per minute. A calorie is the unit that was used in making the measurement; it is the amount of heat required to raise one cubic centimetre of water one degree centigrade. If the heat could be conserved in a layer of water one centimetre thick, the solar radiation would be sufficient to raise the water from freezing-point to boiling-point in fifty minutes. We can be thankful for our protective screen of atmosphere and clouds that cuts off harmful short-wave radiation and reduces heat and light to bearable amounts.

We have to remember that the sun is radiating heat and light in every direction, and that at the distance of the earth this radiation is spread out over a sphere of radius 92·9 million miles. We want the area of this sphere, and we can readily find it from the formula:

$$4\pi r^2 = 4 \times 3·142 \times (92·9 \times 10^6)^2 \text{ square miles}$$
$$= 4 \times 3·142 \times (92·9 \times 10^6 \times 5280 \times 12)^2 \text{ square inches}$$

(To change inches into centimetres we multiply by 2·54.)

$$= 4 \times 3·142 \times (92·9 \times 10^6 \times 5280 \times 12 \times 2·54)^2 \text{ square centimetres}$$
$$= 2·8 \times 10^{27} \text{ square centimetres.}$$

The whole solar radiation, in calories per minute, is twice this number, that is: $5·6 \times 10^{27}$ calories per minute

or 5600 quadrillion calories per minute.

We can turn calories into foot-pounds by using Joule's equivalent: 1 calorie = 3·086 foot-pounds.

$$5·6 \times 10^{27} \text{ calories per minute}$$
$$= 5·6 \times 3·086 \times 10^{27} \text{ foot-pounds per minute}$$
$$= 17·28 \times 10^{27} \text{ foot-pounds per minute}$$

(We can readily turn this into horse-power, since 1 horse-power = 33,000 foot-pounds per minute.)

$$= \frac{17·28 \times 10^{27}}{33,000} \text{ horse-power}$$
$$= ·52 \times 10^{24} \text{ horse-power.}$$

That is, more than half a quadrillion horse-power. If we regard the sun as a great factory for the production of light and heat, the power of the machinery is equivalent to the continuous labour of half a quadrillion horses. Beside this number even the 3½ million horse-power of Niagara seems the merest triviality. For see:

$$\frac{·52 \times 10^{24}}{3½ \times 10^6} = 150,000 \times 10^{12}.$$

So that the sun's amazing output of energy is equal to 150,000 billion Niagaras.

We shall get more comprehensible numbers if we consider the

radiation from, say, a single square inch of the sun's surface. The radius of the sun is 433,000 miles, so the area of its surface is:

$$4\pi r^2 = 4 \times 3 \cdot 142 \times 433{,}000^2 \text{ square miles}$$
$$= 4 \times 3 \cdot 142 \times (433{,}000 \times 5280 \times 12)^2 \text{ square inches}$$
$$= 9 \cdot 46 \times 10^{21} \text{ square inches.}$$

The radiation per square inch is equal to:

$$\frac{\cdot 52 \times 10^{24} \text{ (the total radiation)}}{9 \cdot 46 \times 10^{21} \text{ (number of square inches)}}$$
$$= 55 \text{ horse-power per square inch.}$$

Think of a one-inch square, and of a 55 horse-power engine. The whole power of the engine, transformed completely to heat and light, is poured out continuously through that one-inch square. That is some sort of picture of the enormous output of energy at the sun's surface.

It is interesting to look at the matter from a rather different point of view. We can get the result by considering the fading away of energy according to the inverse square law. We imagine the whole of the energy being radiated from the centre of the sun, and then compare the intensity of radiation at the sun's surface and the earth's surface. We have:

$$\frac{\text{energy at sun's surface}}{\text{energy at earth's surface}} = \frac{(\text{distance of earth})^2}{(\text{distance of sun's surface})^2}$$
$$= \frac{92 \cdot 9^2 \times 10^{12}}{433{,}000^2}$$
$$= 46{,}030.$$

The energy at the sun's surface is about 46,000 times what it is at the earth's surface.

We have seen that the solar energy at the earth's surface:

$$= 2 \text{ calories per square centimetre per minute}$$
$$= 2 \times 3 \cdot 086 \text{ foot-pounds per square centimetre per minute}$$
$$= 2 \times 3 \cdot 086 \times 2 \cdot 54^2 \text{ foot-pounds per square inch per minute}$$
$$= 39 \cdot 8 \text{ foot-pounds per square inch per minute.}$$

Energy at the sun's surface is 46,030 times as great:

$$= 39 \cdot 8 \times 46{,}030 \text{ foot-pounds per square inch per minute}$$
$$= \frac{39 \cdot 8 \times 46{,}030}{33{,}000} \text{ horse-power per square inch}$$
$$= 55 \text{ horse-power per square inch.}$$

Apart from a few factors which cancel in the first method, the arithmetic is the same in both methods.

The vast outpouring of solar energy has been accounted for as the change of mass into energy. That is the only suggestion that comes anywhere near to meeting the difficulty of the enormous size of the energy. We have the equation:

$$e = mc^2.$$

We can use the equation to find the amount of mass that must be converted into energy to keep up the sun's output of radiation. We have to keep to the same units throughout, say foot-pound-second units. The energy will be in foot-poundals per second, the mass in pounds, and the speed of light in feet per second.

We have seen that the sun's output of energy is:

$17\cdot28 \times 10^{27}$ foot-pounds per minute
$= 17\cdot28 \times 32 \times 10^{27}$ foot-poundals per minute
$= 17\cdot28 \times 32 \times 10^{27} \div 60$ foot-poundals per second
$= 32 \times \cdot288 \times 10^{27}$ foot-poundals per second.

The speed of light is:

$186{,}000 \times 5280$ feet per second.

We use these quantities in the equation:

$$e = mc^2.$$

$$32 \times \cdot288 \times 10^{27} = m \times (186{,}000 \times 5280)^2,$$

$$m = \frac{32 \times \cdot288 \times 10^{27}}{186{,}000^2 \times 5280^2}$$

$= 9\cdot56 \times 10^9$ pounds per second (or not far short of $10^{10}$ pounds per second)

$$= \frac{9\cdot56 \times 10^9}{2240} \text{ tons per second}$$

$=$ about $4\frac{1}{4}$ million tons per second.

That is to say, the sun is losing mass at the rate of $4\frac{1}{4}$ million tons per second, and this mass is radiated as light and heat. A year is about $31\cdot56$ million seconds, so that the amount of mass converted into energy in a year is:

$4\frac{1}{4} \times 10^6 \times 31\cdot56 \times 10^6$ tons
$=$ about $134$ billion tons.

That may seem a large and frightening amount, as though the sun were wasting its substance riotously, as indeed it is. It has a lot of substance to waste. The masses of the heavenly bodies appear stupendous when we express them in tons. The mass of the earth is about 6000 trillion tons. So that a mass the size of the earth would be used up by the sun, as a means of radiating energy, in:

$$\frac{6000 \times 10^{18}}{134 \times 10^{12}} \text{ years}$$

$$= \text{about 45 million years.}$$

The whole mass of the sun is 330,000 times the mass of the earth. At its present rate of consumption, one per cent of the sun would be dissipated in:

$$45 \times 3300 \text{ million years}$$
$$= \text{nearly 150,000 million years,}$$

a very handsome lease of life for the sun, and the solar system which depends on it. The 3000 million years or so of the sun's past history fades into insignificance compared with that vast period of time. We could hardly expect the sun to radiate away more than a small fraction of its mass; it would be asking far too much to expect the sun to transform its mass to the last billion tons. The sun will shrink in size, and cool, and probably fade out whilst it still retains a huge mass. So far as we know the nuclei of atoms cannot be completely transformed to energy; in nuclear reactions that have been observed it is only a small fraction of the mass that is changed to energy. But of this we can be pretty sure: if the sun is dying, it is certainly dying a very lingering death, and there is no immediate anxiety about the future.

The earth receives only a very small fraction of the sun's export of energy. At the distance of the earth the whole amount of energy is spread evenly over a sphere of radius 92·9 million miles. The area of this sphere is:

$$4\pi r^2 = 4 \times 3 \cdot 142 \times 92 \cdot 9^2 \times 10^{12} \text{ square miles.}$$

On this sphere we have to think of a comparatively small circle to represent the earth; the curvature is negligibly small. The radius of the circle is 4000 miles, so the area is:

$$\pi r^2 = 3 \cdot 142 \times 4000^2 \text{ square miles.}$$

We want to know what fraction this is of the whole sphere. It is:

$$\frac{3 \cdot 142 \times 4000^2}{4 \times 3 \cdot 142 \times 92 \cdot 9^2 \times 10^{12}} = \text{about } \frac{1}{2,160,000,000},$$

or a little less than 1 part in 2000 millions.

The earth's share of the 134 billion tons of the sun's mass that is converted into energy every year is:

$$\frac{134 \text{ billion tons}}{2000 \text{ millions}} = 67,000 \text{ tons.}$$

Thus the complete consumption of 67,000 tons per annum is sufficient to produce all the heat and light the sun pours on the earth. Compare this with the poor effects obtained by the partial release of the energy of 200 million tons of coal and more, in England alone. Think of the minute patches of illumination on a winter night, and the vast stretches of unlit fields and hills. Think of the inefficient heating in houses, and vast frozen stretches outside. That is what we have to compare with the full blaze of light at noon, and the hottest summer sunshine over the whole country. If it were possible to release nuclear energy, the whole world could be independent of the sun at a mere expenditure of about 100,000 tons annually. So far as actual practice is concerned, that is probably no more than a dream. But it does give a sort of picture of the enormous forces at work in the sun and the stars, and of the enormous amount of energy stored up in the nuclei of atoms.

Since this chapter was written we have had terrifying illustrations of what nuclear energy is capable of in destruction, and an exhilarating promise of what may be possible when this mighty force has been harnessed.

# CHAPTER 10

## *Magnitudes of the Stars*

THE MATHEMATICS of the magnitudes of stars is peculiar and interesting. The peculiarity arises from the fact that modern exact methods of measurement have been superimposed on an ancient vague method.

We classify stars according to their brightness as stars of the first magnitude, stars of the second magnitude, and so on. Until recently the classification was done vaguely by eye estimation. The brightest stars were classified together as first magnitude stars, though they varied considerably in brightness. Stars of the second magnitude had about the brightness of most of the stars of the Plough. The feeblest of the stars forming the Plough, Megrez, is about the third magnitude. And so the classification proceeds down to stars of the sixth magnitude which are the feeblest that can be detected by the unaided eye.

We still keep the old form of classifying the brightness of stars, though it is now possible to measure the brightness much more accurately. The light of a star whose brightness is to be measured is allowed to fall on a photo-electric cell. The cell gives a current proportional to the light which falls on it, and so gives an exact measure of the brightness of the star.

When we begin to measure the brightness of stars exactly, we can no longer be content with the very approximate classification into integral magnitudes; we are driven to the use of fractions. We still keep to the idea that the smaller the magnitude the brighter the star is. Thus a star of magnitude ·9 is a little brighter, and a star of magnitude 1·1 a little dimmer, than a first magnitude star.

The original scheme of magnitudes seems to have been based on the idea that there is nothing brighter than a first magnitude star; apart of course from the sun and the moon. Well there is; and so we are driven back to magnitude 0 as one magnitude brighter than magnitude 1; then to magnitude −1 as one magnitude brighter still; and so on to $-2$, $-3$, and $-4$, and we can continue just as far as we need.

When the modern exact measurement of magnitudes began it was found that a standard first magnitude star was just about 100 times as bright as one of the faint sixth magnitude stars. This relation was adopted as the basis of an exact scale of brightness. In this exact scale a first magnitude star has exactly 100 times the brightness of a sixth magnitude star.

The exact scale proceeds by multiples. The second magnitude is a certain fraction of the first; we will call this fraction $m$. The third magnitude is $m$ times the second, or $m^2$ times the first. The fourth is $m^3$ times the first, the fifth $m^4$ times the first, and the sixth $m^5$ times the first. We know that the sixth is one-hundredth of the first, so:

$$m^5 = \tfrac{1}{100}.$$
$$\log m^5 = \log \tfrac{1}{100},$$
$$5 \log m = -2 = -5 + 3,$$
$$\log m = -1 + \tfrac{3}{5} = \bar{1}\cdot 6,$$
$$m = \text{antilog } \bar{1}\cdot 6$$
$$= \cdot 3981.$$

And so we find that in this scale each magnitude is ·3981 times the previous one, or just about $\tfrac{2}{5}$. The comparative brightness of the first six magnitudes is:

|  |  |  |  |
|---|---|---|---|
| 1st: | 1, | 2nd: | ·3981, |
| 3rd: | $·3981^2 = ·1585$, | 4th: | $·3981^3 = ·0631$, |
| 5th: | $·3981^4 = ·0251$, | 6th: | $·3981^5 = ·01$. |

If we use the approximate value of $\tfrac{2}{5}$ we have:

| | | | | | |
|---|---|---|---|---|---|
| 1st 1, | | 2nd $\tfrac{2}{5} = ·4$, | | 3rd $\tfrac{4}{25} = ·16$, | |
| 4th $\tfrac{8}{125} = ·064$, | 5th $\tfrac{16}{625} = ·0256$, | 6th $\tfrac{32}{3125} = ·01024$. |

So that the differences are not very great.

When we go in the opposite direction, from duller to brighter, we have to multiply by the reciprocal of ·3981.

$$\frac{1}{\cdot 3981} = 2 \cdot 512.$$

So that we have to multiply by 2·512, or a little more than $2\tfrac{1}{2}$. Thus a star of magnitude 0 is 2·512, or just about $2\tfrac{1}{2}$ times brighter than magnitude 1. Here are a few negative magnitudes set down using both 2·512 and $2\frac{1}{2}$:

| Magnitude | 2·512 | $2\frac{1}{2}=2\cdot5$ |
|---|---|---|
| 0 | 2·512 | $2\frac{1}{2}=2\cdot5$ |
| −1 | $2\cdot512^2=6\cdot310$ | $2\cdot5^2=6\cdot25$ |
| −2 | $2\cdot512^3=15\cdot85$ | $2\cdot5^3=15\cdot625$ |
| −3 | $2\cdot512^4=39\cdot81$ | $2\cdot5^4=39\cdot0625$ |
| −4 | $2\cdot512^5=100$ | $2\cdot5^5=97\cdot65625.$ |

One of the penalties for not having a strictly mathematical scale (which would have 0 for zero point instead of 1) is that 2 balances with −1, 3 with −2, and so on. We always have to make allowance for the displaced zero point.

One of the nearest approaches to standard first magnitude brightness is Aldebaran. This is a bright star in the Hyades, the V-shaped group between Orion and the Pleiades. Its magnitude is 1·1, that is ·1 of a magnitude lower than 1, or $\cdot3981^{\cdot1}$. It is the awkward zero that makes it necessary to treat the magnitude as if it were a magnitude of ·1.

$$\log \cdot3981^{\cdot1} = \cdot1 \times \log \cdot3981$$
$$= \cdot1 \times \bar{1}\cdot6$$
$$= \bar{1}\cdot96,$$
$$\text{antilog } \bar{1}\cdot96 = \cdot9120.$$

So that Aldebaran has about ·91 of the brightness of a first magnitude star.

The Belt of Orion points up to Aldebaran; it also points down to Sirius, which is interesting as the brightest of all the fixed stars; the red and blue flashes of Sirius may be seen in the winter sky. Sirius has a magnitude of −1·58, so that it is 2·58 magnitudes brighter than a first magnitude star. It is therefore:

$$2\cdot512^{2\cdot58} \text{ times as bright.}$$
$$\log 2\cdot512^{2\cdot58} = 2\cdot58 \times \log 2\cdot512$$
$$= 2\cdot58 \times \cdot4$$
$$= 1\cdot032,$$
$$\text{antilog } 1\cdot032 = 10\cdot76.$$

So that Sirius is $10\frac{3}{4}$ times as bright as a first magnitude star.

The brightest of the planets is Venus. It varies greatly in brightness, partly because of its varying distance from the earth, and partly because of its phases. When it is brightest its magnitude is $-4$. It is then 5 magnitudes brighter than, that is to say it is 100 times as bright as, a first magnitude star, and 9 or 10 times as bright as Sirius.

The magnitude of the sun, as we should expect, can be expressed by a small number, even though it is so very much brighter than any other celestial object. The reason lies of course in the logarithmic nature of magnitudes: magnitudes are logarithms to base 2·512, so long as we remember that the zero point is magnitude 1 and that they increase in the left (negative) direction. By continually multiplying by 2·512 we soon reach a number large enough to express even the exceeding brightness of the sun.

The magnitude of the sun has been estimated at $-26\cdot7$. That is to say it is 27·7 magnitudes brighter than a first magnitude star, or $2\cdot512^{27\cdot7}$ times as bright. Another way of saying the same thing is to say that the light of the sun is equivalent to that of $2\cdot512^{27\cdot7}$ first magnitude stars. Let us find this number.

$$\log 2\cdot512^{27\cdot7} = 27\cdot7 \log 2\cdot512$$
$$= 27\cdot7 \times \cdot4$$
$$= 11\cdot08,$$
$$\text{antilog } 11\cdot08 = 1\cdot202 \times 10^{11}$$
$$\text{or } \cdot1202 \times 10^{12}.$$

So that the sun has the brightness of about an eighth of a billion, or 120,000 million first magnitude stars.

A number of comparisons have been made of the relative brightness of the sun and the moon. The estimates vary considerably, the most recent is about 240,000. Let us see how many magnitudes this represents. $x$ magnitudes stands for a factor of $2\cdot512^x$, and this is to equal 240,000.

$$2\cdot512^x = 240,000,$$
$$\log 2\cdot512^x = \log 240,000$$
$$x \log 2\cdot512 = 5\cdot3802,$$
$$x = \frac{5\cdot3802}{\cdot4}$$
$$= 13\cdot45.$$

So that the moon is 13·45 magnitudes lower than the sun.

Because of the system of counting backwards for magnitudes we have to add 13·45 to the sun's magnitude.

$$\text{Magnitude of moon} = -26\cdot7 + 13\cdot45$$
$$= -13\cdot25.$$

The magnitude of the full moon is probably about $-13\frac{1}{4}$.

What is the magnitude of the sun as seen from Neptune, on the outskirts of the Solar System?

Neptune's distance from the sun is almost exactly 30 times that of the earth, so that the lighting effect is reduced to:

$$\frac{1}{30^2} = \frac{1}{900}.$$

We change 900 to a magnitude:

$$2\cdot512^x = 900,$$
$$x \log 2\cdot512 = \log 900 = 2\cdot9542,$$
$$x = \frac{2\cdot9542}{\cdot4} = 7\cdot385 \text{ or } 7\cdot4.$$

So the magnitude of the sun as seen from Neptune is:

$$-26\cdot7 + 7\cdot4 = -19\cdot3.$$

That is, $19\cdot3 - 13\cdot25 =$ about 6 magnitudes up on the full moon. The sun as seen from Neptune is $2\cdot512^6 = 250$ times as bright as the full moon. It would take more than 250 full moons to supply the same amount of light to Neptune as does the diminutive sun, which is only $\frac{1}{30}$ of the width of the sun as seen from the earth. And that is another picture of the power of the sun.

250 is actually an underestimate; the approximations have erred on that side. A closer approximation is:

$$\frac{240,000}{900} = \text{nearly 270 full moons.}$$

The apparent brightness of a star, that is its magnitude, may be affected by two quite different causes. In the first place a star may be intrinsically bright or dull. And in addition the apparent magnitude is affected by the distance of the star from the earth. If any star were removed to twice its present distance from the earth, its apparent brightness would be reduced to a quarter. Thus we may have comparatively dull stars that appear very bright because of their nearness to the earth; and intrinsically bright stars that appear faint because of their remoteness.

In order to have a comparison of the actual brightness of stars we imagine them placed at equal distances from the earth, and calculate what their magnitudes would be if placed at that distance. Magnitudes calculated in this way are called absolute magnitudes.

The standard distance chosen for calculating absolute magnitudes is 10 parsecs, that is 10 times the distance at which there is a parallax of one second of arc, or the distance at which there is a parallax of one-tenth of a second.

10 parsecs is equal to:

$$\frac{\text{distance of sun}}{\sin 0 \cdot 1''} = \frac{92 \cdot 9 \times 10^6}{\cdot 0000004848} \text{ miles}$$

$$= 192 \text{ billion miles.}$$

We cannot of course find the absolute magnitude of a star unless its parallax has been measured. We have the requisite data for Sirius. Its apparent magnitude is $-1 \cdot 58$; and its distance is $2 \cdot 70$ parsecs; we can therefore calculate the absolute magnitude of Sirius.

If Sirius were removed to a distance of 10 parsecs, its brightness would be reduced to:

$$\frac{2 \cdot 7^2}{10^2} = \cdot 0729 \text{ of its present value.}$$

Now the magnitude of $-1 \cdot 58$ is $2 \cdot 58$ magnitudes up on a first magnitude star, and:

$$2 \cdot 512^{2 \cdot 58} = 10 \cdot 76,$$

so that Sirius is $10 \cdot 76$, or about $10\frac{3}{4}$ times as bright as the standard first magnitude star.

At the standard distance of 10 parsecs this brightness is reduced to:

$$10 \cdot 76 \times \cdot 0729 = \cdot 7844.$$

That is to say, at the standard distance Sirius has a brightness equal to $\cdot 7844$ times that of a first magnitude star. We now have to change this fraction into a magnitude.

$$2 \cdot 512^x = \cdot 7844,$$
$$x \log 2 \cdot 512 = \log \cdot 7844,$$
$$x = \frac{\overline{1} \cdot 8945}{\cdot 4}$$
$$= - \cdot 26.$$

Hence Sirius is ·26 of a magnitude below 1. That is, its absolute magnitude is 1·26.

Canopus, the bright star in Argus, is interesting on account of its intense intrinsic brightness. It has an apparent magnitude of − ·86, although it is at the considerable distance of 200 parsecs. At the standard distance of 10 parsecs its brightness would be increased by a factor of:

$$\frac{200^2}{10^2} = 20^2 = 400.$$

We have to change this factor into a magnitude.

$$2\cdot512^x = 400,$$
$$x = \frac{\log 400}{\log 2\cdot512} = \text{about } 6\cdot5.$$

We have to subtract 6·5; that is:

$$-\cdot86 - 6\cdot5 = -7\cdot36,$$

and that is the absolute magnitude.

If Canopus were to move up to the standard distance of 10 parsecs, its brightness would be:

$2\cdot512^{8\cdot36} = $ about 2200 times that of a first magnitude star, or 22 times as bright as Venus at its brightest.

Let us see how the sun comes out in this comparison of absolute magnitudes. At the standard distance, the square of the sun's distance from the earth would be:

$$\left(\frac{192 \times 10^{12}}{92\cdot9 \times 10^6}\right)^2 = 4\cdot27 \times 10^{12} \text{ times as great as it actually is.}$$

We change this into a magnitude:

$$2\cdot512^x = 4\cdot27 \times 10^{12},$$
$$x = \frac{\log (4\cdot27 \times 10^{12})}{\log 2\cdot512}$$
$$= 31\cdot57.$$

We have to add this magnitude:

$$-26\cdot7 + 31\cdot57 = 4\cdot87.$$

So that viewed from the standard distance of 10 parsecs, the sun would be a little higher than a fifth magnitude star, and not much brighter than the limit of what is visible to the naked eye.

# CHAPTER 11

## Symbols

ONE OF the things that frighten people away from the simple and straightforward subject of mathematics is the symbols that are used. Every branch of mathematics has its own set of symbols and we cannot get very far without them. It really is extraordinary the fear that many people have of symbols; except of course the symbols they are so familiar with, that they have ceased to regard them as symbols. I have seen people frozen stiff at the mere sight of $\pi$, as though Medusa had suddenly peeped at them. Whereas another symbol, $3\frac{1}{7}$ or even $3\cdot14159$, would have left them with usable wits.

Perhaps the most valuable improvement in mathematics was the introduction of the Arabic symbols for numbers. The improvement on the older system of symbols was so great that a multiplication sum that would have taxed the skill of a great mathematician can now be performed with ease by a schoolboy.

Suppose we want to find four thousand eight hundred and seventy-six multiplied by nine thousand five hundred and two. We could say:

Nine thousand times four thousand is thirty-six millions; nine thousand times eight hundred is seventy-two hundred thousand, or seven million two hundred thousand; so far that is forty-three millions two hundred thousand. Nine thousand times seventy is six hundred and thirty thousand; so we now have forty-three millions eight hundred and thirty thousand. Nine thousand times six is fifty-four thousand; making forty-three millions eight hundred and eighty-four thousand. Five hundred times four thousand is twenty hundred thousand, or two millions; making forty-five millions eight hundred and eighty-four thousand. Five hundred times eight hundred is forty ten thousands, or four hundred thousand; making forty-five millions twelve hundred thousand, and eighty-four thousand; or forty-six millions two hundred and eighty-four thousand. Five hundred times seventy is thirty-five thousand; making forty-six millions, three hundred and nineteen thousand. Five hundred times six is three thousand; making forty-six millions three hundred and twenty-two thousand. Twice four thousand is eight thousand; making forty-six millions three hundred and thirty thousand. Twice eight hundred is one thousand six hundred; making forty-six million three hundred and thirty-one thousand, six hundred. Twice seventy is one hundred and forty, and twice six is twelve; making one hundred and fifty-two; and a grand total of forty-six millions three hundred and thirty-one thousand seven hundred and fifty-two.

Using symbols, a schoolboy would set that down:

$$
\begin{array}{r}
4876 \times \\
9502 \\
\hline
43884000 \\
2438000 \\
9752 \\
\hline
46331752.
\end{array}
$$

There can be no doubt about the convenience and added certainty of using symbols.

Some of the symbols we use are mere contractions, like £ s. d. for pounds, shillings and pence; = is a shorthand way of writing 'equals', and ≡ of writing 'is identical with'. Other symbols are what are called *operators*, that is, they are instructions to carry out some process. + is an instruction to add the quantities between which it is written; − is an instruction to subtract the quantity which follows it from the quantity which precedes it. +, −, × and ÷ are the most commonly used operators. $\sqrt{\phantom{x}}$ may be read as an instruction to find the square root of a number, and $\sqrt[3]{\phantom{x}}$ as an instruction to find the cube root.

Many of the symbols used in algebra have a double purpose. They are a convenient, shorthand way of writing down results, so that the relations may be readily seen. And in addition to that the symbols generalise the results, so that they are true for all numbers.

We can say that the difference between the squares of two numbers is equal to the sum of the numbers multiplied by their difference. In symbols we write:

$$x^2 - y^2 = (x+y)\ (x-y).$$

In the symbolic form the statement is obviously true, since we have only to multiply the two factors to prove it. And having proved the proposition in general we know that it is true for any particular case. We know for example that:

$$(83\tfrac{5}{8})^2 - (67\tfrac{3}{4})^2 = (83\tfrac{5}{8} + 67\tfrac{3}{4})\ (83\tfrac{5}{8} - 67\tfrac{3}{4}),$$

without going to the trouble of working out the squares.

Here is an example of the kind of problem that figures in elementary algebra:

A lawn is 32 yards by 84 yards. Round it there is a path the same width on all sides. The area of the lawn is $\tfrac{7}{8}$ of the outside area. What is the width of the path?

The outside width is 32 yards added to twice the width of the path, and the outside length is 84 yards added to twice the width of the path. So we have:

32 times 84 square yards is equal to $\tfrac{7}{8}$ of 32 yards added to twice the width of the path, and this length multiplied by 84 yards added to twice the width of the path.

The multiplication is: 32 times 84, added to 32 times twice the width of the path, added to 84 times twice the width of the path, added to twice the width of the path times twice the width of the path. And this equals: 32 times 84, added to 232 times the width of the path, added to 4 times the square of the width of the path.

We thus arrive at the equation:

32 times 84 is equal to $\tfrac{7}{8}$ of 32 times 84, added to $\tfrac{7}{8}$ of 232 times the width of the path, added to $\tfrac{7}{8}$ of 4 times the square of the width of the path.

Almost any one might be excused for sticking at that point. And yet, with symbols, any one who knows about quadratics can

readily solve the problem. Instead of the cumbrous statement we have:
$$32 \times 84 = \tfrac{7}{8}(32 + 2x)(84 + 2x).$$

$$\therefore \ \tfrac{8}{7} \times 32 \times 84 = 32 \times 84 + x(2 \times 32 + 2 \times 84) + 4x^2,$$

$$\therefore \ 4x^2 + 232x - 384 = 0,$$

or
$$x^2 + 58x - 96 = 0.$$

$$x = \frac{-58 \pm \sqrt{58^2 + 4 \times 96}}{2}$$

$$= \frac{-58 \pm 61\cdot2}{2}$$

$$= 1\cdot6 \text{ or } -59\cdot6.$$

The answer we want is 1·6 yards.

There can be no reasonable doubt about the value of symbols, or of the clarity and simplicity they introduce into mathematical operations. An attempt to do any kind of advanced work without using the appropriate symbols gets one into hopeless tangles.

The chief use of symbols is to simplify processes. They only become mystifying when we have not taken the trouble to understand the simple things they mean.

We use the symbol $\lfloor n$ or $n!$ to mean

$$n \times (n-1) \times (n-2) \times \ldots \times 3 \times 2 \times 1.$$

This quantity is called 'factorial $n$'. Factorial 6, for example, is $6 \times 5 \times 4 \times 3 \times 2 \times 1$.

We can readily show that the number of combinations of 8 things, taking 3 at a time, is:

$$\frac{8 \times 7 \times 6}{3 \times 2 \times 1}.$$

We can choose any of the 8 for the first, any of the remaining 7 for the second; and we can associate each of the first with each of the second. So we have $8 \times 7$ combinations altogether. Then we can choose any of the remaining 6 for the third, and we can associate each of these with each of the $8 \times 7$. And so we obtain $8 \times 7 \times 6$ combinations. But these are not all different. If we take the eight letters, $A$ to $H$, we should include: $ABC$, $ACB$, $BAC$, $BCA$, $CAB$, $CBA$; and so on for each other combination of 3 letters. Now the number of these combinations is $3 \times 2 \times 1$, or $\lfloor 3$.

(We get at that by the same process of reasoning.) $8 \times 7 \times 6$ is therefore $\lfloor 3$ times too big, and so the number of combinations is:

$$\frac{8 \times 7 \times 6}{\lfloor 3}.$$

We write $^8C_3$ for combinations of 8 things taken 3 at a time. We can use this symbol with confidence, so long as we keep clearly in mind exactly what it means.

$$^{12}C_5 = \frac{12 \times 11 \times 10 \times 9 \times 8 \text{ (5 factors)}}{\lfloor 5},$$

$$^nC_m = \frac{n(n-1)(n-2) \dots \text{ to } m \text{ factors}}{\lfloor m}.$$

(If $m=2$ the final term is $n-1$; if $m=3$, the final term is $n-2$, and so on. So the final term is $n-m+1$.)

$$= \frac{n(n-1)(n-2) \dots (n-m+1)}{\lfloor m}.$$

The symbol $^nC_m$ gives a very simple way of writing the binomial theorem. We know that:

$$(x+a)^n = x^n + nx^{n-1}a + \frac{n(n-1)}{\lfloor 2} x^{n-2}a^2 + \frac{n(n-1)(n-2)}{\lfloor 3} x^{n-3}a^3 + \text{etc.}$$
$$= x^n + {^nC_1}x^{n-1}a + {^nC_2}x^{n-2}a^2 + {^nC_3}x^{n-3}a^3 + \text{etc.}$$

We can use that form to write down any term we happen to want. Suppose we want the sixth term of $(x+a)^9$. This term is:

$$^9C_5 x^{9-5}a^5 = \frac{9 \times 8 \times 7 \times 6 \times 5}{\lfloor 5} x^4 a^5$$
$$= 126 x^4 a^5.$$

We may notice in passing that the coefficient of $x^n$ should be $^nC_0$, so that $^nC_0 = 1$. And $^nC_n = 1$. (There is only one way in which we can choose $n$ things out of $n$.)

Now let us see how the symbol $^nC_m$ can be used to establish a general proposition.

Suppose we write $x = a = 1$. Then $(x+a)^n = 2^n$. So:

$$2^n = {^nC_0} + {^nC_1} + {^nC_2} + \dots + {^nC_n}.$$

Now put $x = 1$ and $a = -1$. Then $(x+a)^n = 0$. So:

$$0 = {^nC_0} - {^nC_1} + {^nC_2} - \dots (-1)^n {^nC_n}.$$

(Since the series starts with $^nC_0$, and not $^nC_1$, we have $^nC_n$ as the

$(n+1)$th term. The sign is given by $(-1)^n$, that is 1 less than the term, $(-1)^1$ for the second term, $(-1)^2$ for the third term, and $(-1)^{n+1-1} = (-1)^n$ for the $(n+1)$th term.)

Now add the two results:

$$2^n = 2\,({}^nC_0 + {}^nC_2 + {}^nC_4 + \ldots).$$

We do not know whether to include ${}^nC_n$, because we do not know whether its sign is $+$ or $-$. We can get over this difficulty by using $2n$ instead of $n$.

$$2^{2n} = 2\,({}^{2n}C_0 + {}^{2n}C_2 + {}^{2n}C_4 + \ldots + {}^{2n}C_{2n}).$$

We include the last term because its sign is given by $(-1)^{2n} = +1$.

Instead of adding we can subtract.

$$2^{2n} = 2\,({}^{2n}C_1 + {}^{2n}C_3 + {}^{2n}C_5 + \ldots + {}^{2n}C_{2n-1}).$$

And so we arrive at the conclusion that the sum of the odd coefficients is equal to the sum of the even coefficients. We have proved this for even powers $(2n)$. It is just as easy to prove it for odd powers; we have only to assume that $n$ is odd, and therefore $(-1)^n = -1$.

People are sometimes worried by the symbol $f(x)$, which is used to stand for any function of $x$. A function of $x$ is any mathematical expression that includes $x$: $x^2 + ax + b$, $3^x$, $\sin x$, $\sin^3 x \cos^2 x$, and so on. We use the symbol $f(x)$ when we do not, for the moment, wish to specify a particular function; we use it because it gives results in a simpler and also in a more general form.

We extend the meaning of the symbol to include $f(x, y)$; that is a function of both $x$ and $y$, such as $x^y$ or $\sin x \cos y$. With this notation $f(1)$ stands for the function we are considering, with 1 substituted for $x$; $f(2)$ stands for the same function with 2 substituted for $x$; $f(0)$ stands for the same function with 0 substituted for $x$.

Thus if:
$$f(x) = x^2 + bx + c,$$
$$f(1) = 1 + b + c,$$
$$f(3) = 9 + 3b + c,$$
$$f(0) = c,$$
$$f(t) = t^2 + bt + c.$$

If we are dealing with two or more functions of $x$, we can call them $f_1(x)$, $f_2(x)$, and so on (the $f_1$ function, the $f_2$ function, and so on).

Another symbol which sometimes worries people is $\frac{d}{dx}$, which is used in the differential calculus. It is read as a single symbol, and we have to resist the very elementary temptation to cancel the $d$'s. $\frac{dy}{dx}$ is read '$d, y$ by $d, x$', and it is a definite quantity produced in a definite way.

$y$ is a function of $x$, say, $y = f(x)$.

Suppose we increase $x$ by a small quantity which we will call $\Delta x$; $\Delta x$ is a single symbol for this small quantity. When we increase $x$ we also increase $y$, by a different amount which we will call $\Delta y$. Then:

$$y + \Delta y = f(x + \Delta x).$$

That is, we increase $x$ to $x + \Delta x$, and so we increase $y$ to $y + \Delta y$.

Now subtract $y = f(x)$.

We have: $\Delta y = f(x + \Delta x) - f(x)$.

Divide both sides by $\Delta x$:

$$\frac{\Delta y}{\Delta x} = \frac{f(x + \Delta x) - f(x)}{\Delta x}.$$

That is true no matter how small $\Delta x$ may be. When $\Delta x$ is made extremely small $\Delta y$ also is extremely small. We may, and do, get a definite limit toward which $\frac{\Delta y}{\Delta x}$ approaches, when $\Delta x$, and therefore $\Delta y$ also, approach zero. We call this limit $\frac{dy}{dx}$.

$$\frac{dy}{dx} = \mathrm{Lt.}_{\Delta x \to 0} \frac{f(x + \Delta x) - f(x)}{\Delta x}.$$

$\frac{dy}{dx}$ has that definite meaning; it is called the *derivative* of $y$ or $f(x)$.

Difficulties sometimes arise in evaluating the limit, but the statement of the process is simple and straightforward.

The simplest case is when $y$ is a power of $x$, say $x^3$.

$$\frac{dy}{dx} = \mathrm{Lt.}_{\Delta x \to 0} \frac{(x + \Delta x)^3 - x^3}{\Delta x}$$

$$= \mathrm{Lt.}_{\Delta x \to 0} \frac{3x^2 \cdot \Delta x + 3x(\Delta x)^2 + (\Delta x)^3}{\Delta x}$$

$$= \mathrm{Lt.}_{\Delta x \to 0} 3x^2 + \Delta x(3x + \Delta x)$$

$$= 3x^2.$$

Here is an example of a simple artifice used in finding a derivative:

$$y = \sin x,$$

$$\frac{dy}{dx} = \text{Lt.}_{\Delta x \to 0} \frac{\sin (x + \Delta x) - \sin x}{\Delta x}.$$

We have a formula for the difference of two sines, and we use this:

$$\sin A - \sin B = 2 \cos \frac{A + B}{2} \sin \frac{A - B}{2}.$$

For $A$ we have $x + \Delta x$ and for $B$ we have $x$. So $\dfrac{A + B}{2} = x + \frac{1}{2}\Delta x$ and $\dfrac{A - B}{2} = \frac{1}{2}\Delta x$.

$$\frac{dy}{dx} = \text{Lt.}_{\Delta x \to 0} \frac{2 \cos (x + \frac{1}{2}\Delta x) \sin \frac{1}{2}\Delta x}{\Delta x}$$

$$= \text{Lt.}_{\Delta x \to 0} \cos (x + \frac{1}{2}\Delta x) \frac{\sin \frac{1}{2}\Delta x}{\frac{1}{2}\Delta x}$$

$$= \cos x.$$

(We know that for very small angles the sine is equal to the angle. So the limit of $\dfrac{\sin \frac{1}{2}\Delta x}{\frac{1}{2}\Delta x} = 1$, when $\Delta x$ approaches o.)

That is,
$$\frac{d \sin x}{dx} = \cos x.$$

$\int$ is the symbol for integration, which is the inverse of differentiation. $\int y \, dx$ is an instruction to find the quantity whose derivative is $y$. We know, for example, that $\dfrac{d \sin x}{dx} = \cos x$, so:

$$\int \cos x \, dx = \sin x.$$

That is, $\sin x$ is the integral of $\cos x$, or the quantity whose derivative is $\cos x$.

Integration is essentially a more difficult process than differentiation. But the difficulty is not in the symbolism. It lies in the fact that we have to hunt for integrals by comparison with known derivatives. And we can never be sure that there is an integral to be found, just as we cannot be sure that an expression, written at random, will factorise.

Here is another simple example to show how integrals are found by comparison. We know that $\frac{dx^n}{dx} = nx^{n-1}$, and we want to find $\int x^n dx$.

By comparison we see that $x^n$ is the integral of $nx^{n-1}$; so the integral of $x^{n-1}$ is $\frac{x^n}{n}$. We now write $n+1$ in place of $n$, so that $n-1$ becomes $n$.

$$\int x^n \, dx = \frac{x^{n+1}}{n+1},$$

and we can readily check that by differentiating.

Taylor's Theorem is a very striking example of the advantage of using symbols. We can write this theorem:

$$f(x+h) = f(x) + hf'(x) + \frac{h^2}{\underline{|2}} f''(x) + \dots + \frac{h^p}{\underline{|p}} f^p(x+\theta h).$$

$f(x+h)$ and $f(x)$ we already know the meaning of. $f'(x)$ is another way of writing the derivative of $f(x)$; $f''(x)$ is the second derivative, that is, the derivative of the derivative. $\theta$ is a positive proper fraction, so that it is less than 1.

Taylor's Theorem is true so long as $f(x)$ and all its derivatives are finite and continuous. We should get very odd results if one or more of the derivatives suddenly bumped up to infinity, or had a gap in it.

We could go on writing more and more terms of the series, so long as the terms are finite and continuous. The end term is $\frac{h^p}{\underline{|p}} f^p(x+\theta h)$; if this term vanishes when $p$ becomes extremely great, we can continue the terms indefinitely.

There is another form of Taylor's Theorem which is known as Maclaurin's Theorem. We can get it from Taylor's Theorem by the simple device of writing $x$ for $h$, and $o$ for $x$.

$$f(x) = f(o) + xf'(o) + \frac{x^2}{\underline{|2}} f''(o) + \dots + \frac{x^p}{\underline{|p}} f^p(\theta x).$$

$f(o)$ is $f(x)$ with $o$ written for $x$; $f'(o)$ is the first derivative of $f(x)$, with $o$ written for $x$, after the differentiating has been done; and so on.

One of the simplest series that can be got from Maclaurin's Theorem is the series for $e^x$. We know that $\dfrac{de^x}{dx} = e^x$; so all the successive derivatives of $e^x$ are equal to $e^x$. $f(0) = e^0 = 1$; $f'(0) = e^0 = 1$; and so on. So we have:

$$e^x = 1 + x + \frac{x^2}{\underline{2}} + \dots.$$

Now $x$ is a finite quantity, and $\theta x$ is a fraction of $x$. So $f^p(\theta x) = e^{\theta x}$ = a finite quantity. $\dfrac{x^p}{\underline{p}}$ can be made indefinitely small by taking $p$ very great. It can be written:

$$\frac{p}{1} \times \frac{p}{2} \times \frac{p}{3} \times \frac{p}{4} \times \dots.$$

We are multiplying by smaller and smaller fractions, and so ultimately approach zero.

So we can say that:

$$e^x = 1 + x + \frac{x^2}{\underline{2}} + \frac{x^3}{\underline{3}} + \dots \text{ to infinity.}$$

When we write $x = 1$, we have:

$$e = 1 + 1 + \frac{1}{\underline{2}} + \frac{1}{\underline{3}} + \dots \text{ to infinity.}$$

Another simple example is the expansion of $\log(1+x)$. We need to know the derivatives.

$$\frac{d\log(1+x)}{dx} = \frac{1}{1+x}; \quad \frac{d\,\dfrac{1}{1+x}}{dx} = -\frac{1}{(1+x)^2};$$

$$\frac{d-\dfrac{1}{(1+x)^2}}{dx} = \frac{2}{(1+x)^3}; \quad \frac{d\,\dfrac{2}{(1+x)^3}}{dx} = -\frac{2 \cdot 3}{(1+x)^4}; \text{ and so on.}$$

We now have to write $x = 0$.

$$f(x) = \log(1+x), \qquad\qquad f(0) = \log 1 = 0,$$

$$f'(x) = \frac{1}{1+x}, \qquad\qquad f'(0) = \frac{1}{1} = 1,$$

$$f''(x) = -\frac{1}{(1+x)^2}, \qquad\qquad f''(0) = -\frac{1}{1^2} = -1.$$

Continuing, we get: $f'''(0) = \underline{2}, f''''(0) = -\underline{3}$, and so on.

We fill in these values and get:

$$\log (1+x) = f(0) + xf'(0) + \frac{x^2}{\underline{|2}} f''(0) + \dots$$

$$= 0 + x - \frac{x^2}{\underline{|2}} + \frac{x^3}{\underline{|3}} \times \underline{|2} - \frac{x^4}{\underline{|4}} \times \underline{|3} + \text{ etc.}$$

$$= x - \frac{x^2}{2} + \frac{x^3}{3} - \frac{x^4}{4} + \text{etc.}$$

We still have to make sure of the end term. It is $\frac{x^p}{p} \log \theta x$, so that it can be made as small as we like by making $p$ sufficiently great.

Here is another example of the power of symbols. We imagine $\sin x$ expanded as a series of powers of $x$.

$$\sin x \equiv ax + bx^3 + cx^5 + dx^7 + \text{etc.}$$

The reason we have only odd powers of $x$ is this. $\sin x = -\sin (-x)$; so if we write $-x$ for $x$, we should get the same result but with the minus sign. This is only the case with odd powers.

$$\sin (-x) = -ax - bx^3 - cx^5 - dx^7 - \text{etc.}$$
$$= -(ax + bx^3 + cx^5 + dx^7 + \text{etc.}).$$

Even powers would remain unchanged, so we should get a different result if we put in even powers.

Now we differentiate both sides of the identical equation we have assumed. On the left we have $\cos x$, the derivative of $\sin x$. On the right we have: $a + 3bx^2 + 5cx^4 + 7dx^6 + \text{etc.}$ So:

$$\left. \begin{array}{l} \sin x = ax + bx^3 + cx^5 + dx^7 + \text{etc.} \\ \cos x = a + 3bx^2 + 5cx^4 + 7dx^6 + \text{etc.} \end{array} \right\} \quad \text{(i)}$$

Differentiate again and again.

$$\left. \begin{array}{l} -\sin x = 3.2bx + 5.4cx^3 + 7.6dx^5 + \text{etc.} \\ -\cos x = 3.2b + 5.4.3cx^2 + 7.6.5dx^4 + \text{etc.} \end{array} \right\} \quad \text{(ii)}$$

$$\left. \begin{array}{l} \sin x = 5.4.3.2cx + 7.6.5.4dx^3 + \text{etc.} \\ \cos x = 5.4.3.2c + 7.6.5.4.3dx^2 + \text{etc.} \end{array} \right\} \quad \text{(iii)}$$

Now $\cos 0 = 1$. We write $x = 0$ in equations (i), (ii), (iii), etc.

(i)   We get $1 = a$, so we write $1$ for $a$ in the equation we assumed.

(ii)   $-1 = \underline{|3}b$, so we write $-\dfrac{1}{\underline{|3}}$ for $b$.

(iii)   $1 = \underline{|5}c$, so we write $\dfrac{1}{\underline{|5}}$ for $c$.

And so we find that:

$$\sin x = x - \frac{x^3}{\underline{|3}} + \frac{x^5}{\underline{|5}} - \frac{x^7}{\underline{|7}} + \frac{x^9}{\underline{|9}} - \text{etc.}$$

This series is convergent, and so we can use it to find the value of $\sin x$. $x$ is as usual the circular measure of the angle (in radians).

We can get a similar series for $\cos x$ by assuming:

$$\cos x = a + bx^2 + cx^4 + dx^6 + \text{etc.}$$

This series is:

$$\cos x = 1 - \frac{x^2}{\underline{|2}} + \frac{x^4}{\underline{|4}} - \frac{x^6}{\underline{|6}} + \frac{x^8}{\underline{|8}} - \text{etc.}$$

# CHAPTER 12

## *Squaring the Circle*

PEOPLE TALK glibly about 'squaring the circle', and they send methods of achieving it to mathematical papers, without ever having understood what the problem is. The problem is to construct a square or oblong equal in area to the area of a circle. It is a purely intellectual problem; its practical importance is insignificant. The construction has to be carried out according to the methods of Euclidean geometry; and therein lies the snag.

Practically, there is little difficulty about the problem. We have to construct an oblong, length equal to half the circumference, and width equal to the radius. What difficulty there is, is in drawing a straight line equal to the half circumference; that is indeed the root problem of squaring the circle. There are several well-known methods of finding approximate solutions geometrically. One of the neatest and simplest is this:

We draw the diameter $AB$, and the tangent $CD$ at right angles to it. We make angle $BOC = 30°$. We measure $CD = 3$ times the radius. Finally we join $AD$, and $AD$ is a very close approximation to half the circumference.

It is interesting to see how close the approximation is. We call the radius $r$.

$OB = r$, $AB = 2r$, $CD = 3r$.

We want the length $CB$ in terms of $r$. Triangle $BOC$ is half an equilateral triangle, so $OC = 2CB$.

$$OC^2 = CB^2 + OB^2,$$
$$(2CB)^2 = CB^2 + OB^2,$$
or
$$4CB^2 = CB^2 + r^2,$$
$$3CB^2 = r^2,$$
$$CB = \frac{r}{\sqrt{3}}.$$

We start again with the length $AD$, which we want.

$$AD^2 = AB^2 + BD^2 = (2r)^2 + BD^2$$
$$= 4r^2 + (CD - CB)^2$$
$$= 4r^2 + \left(3r - \frac{r}{\sqrt{3}}\right)^2$$
$$= r^2\left\{4 + \left(3 - \frac{1}{\sqrt{3}}\right)^2\right\}$$
$$= r^2\left(4 + 9 + \tfrac{1}{3} - \frac{6}{\sqrt{3}}\right)$$
$$= r^2\left(13\tfrac{1}{3} - 2\sqrt{3}\right)$$
$$= 9 \cdot 86923 r^2,$$

and $\qquad AD = 3 \cdot 14153 r.$

That is an extremely close approximation, but it is not a solution of the intellectual problem of squaring the circle—nor does it profess to be.

Any fraction can be represented geometrically by a straight line. If the fraction had large numerator and denominator, the representation might be extremely difficult; but that is not to the point. So the problem of squaring the circle resolves itself into finding a fraction that is equivalent to $\pi$, the ratio between the circumference and diameter of a circle. I once heard a member of the Brains Trust come very near to squaring the circle. He was answering a question about the people who profess to believe that the value of $\pi$ is $3 \cdot 1416$ exactly. His reply began: 'They say it is not a non-recurring, I mean a recurring decimal, but $3 \cdot 1416$ exactly.' Now if $\pi$ were a recurring decimal its value could be represented by a fraction, and so it could be represented geometrically. The circle would be squared. Actually, $\pi$ is both non-recurring and non-terminating. That is the whole trouble about it: it goes on and on without end and without recurring. It is, as we say, incommensurable; no fraction, common or decimal, can represent the exact ratio of the circumference of a circle to the diameter.

The ancient world seems to have been easily content with a very rough approximation to the value of $\pi$. There is a famous passage in the description of Solomon's Temple (I Kings vii. 23) which runs:

'And he made a molten sea, ten cubits from the one brim to the other: it was round all about, and his height was five cubits: and a line of thirty cubits did compass it round about.'

The dimensions are repeated in II Chronicles iv. 2. It is odd that so rough an approximation as 3 should be used, because an ordinary school measurement—rolling a circle along a measured line—will give a much better result. It is easy enough to establish that the circumference is, very nearly, $3\frac{1}{7}$ times the diameter.

It is useless to rely on measurement when we want great accuracy. We have to resort to calculation.

Archimedes had the idea of inscribing a polygon in a circle and comparing the circumference of the circle with the perimeter of the polygon.

We begin with a square; and we call the radius of the circle 1, that is, we use it as a unit. We want the half perimeter of the square (the half, because we are using the radius and not the diameter).

$AB = \dfrac{1}{\sqrt{2}}$, so the semi-perimeter is $\dfrac{4}{\sqrt{2}} = 2\sqrt{2}$. That is half the number of sides multiplied by $\sqrt{2}$.

$CB$ is the side of an inscribed octagon. We want to find the length of $CB$.

$$
\begin{aligned}
CB^2 &= CA^2 + AB^2 \\
&= CA^2 + CA \cdot AD \\
&= CA(CA + AD) \text{ or } CA \cdot CD \\
&= 2 \times CA \text{ (since } CD = 2 \text{ units)} \\
&= 2(OC - OA) \\
&= 2\left(1 - \dfrac{1}{\sqrt{2}}\right) \\
&= 2 - \sqrt{2}, \\
CB &= \sqrt{2 - \sqrt{2}}.
\end{aligned}
$$

And the semi-perimeter is $4\sqrt{2 - \sqrt{2}}$.

*EB* is the side of a 16-sided polygon, and we want to find its length.

$$EB^2 = EF^2 + BF^2$$
$$= EF^2 + EF.FG$$
$$= EF\,(EF + FG) \text{ or } EF.EG$$
$$= 2 \times EF$$
$$= 2\,(OE - OF) \text{ or } 2\,(1 - OF),$$
$$OF^2 = OB^2 - BF^2$$
$$= 1 - \tfrac{1}{4}\,(2 - \sqrt{2})$$
$$= \frac{1}{2} + \frac{\sqrt{2}}{4} \text{ or } \frac{2 + \sqrt{2}}{4},$$
$$OF = \frac{\sqrt{2 + \sqrt{2}}}{2}.$$

Now we have:
$$EB^2 = 2\,(1 - OF) = 2\left(1 - \frac{\sqrt{2 + \sqrt{2}}}{2}\right)$$
$$= 2 - \sqrt{2 + \sqrt{2}}.$$
$$EB = \sqrt{2 - \sqrt{2 + \sqrt{2}}}.$$

The semi-perimeter is $8\sqrt{2 - \sqrt{2 + \sqrt{2}}}$.

Proceeding in this way we find for a 32-sided polygon:

$$\text{semi-perimeter} = 16\sqrt{2 - \sqrt{2 + \sqrt{2 + \sqrt{2}}}}.$$

And so we can proceed, just as far as we care to go.

In calculating we proceed as follows: find the square root of 2, add 2, find the square root, add 2, find the square root, and so on. Then we subtract from 2, find the square root, and multiply by 2 to the power of the number of square roots we have extracted.

A figure with 32,768 sides, that is $2^{15}$, requires the extraction of 14 square roots, and a final multiplication by 16,384. This gives the value of $\pi$ to seven decimal places. $\pi = 3.1415926$.

We could use Pythagoras' method to find the value of $\pi$ to just as many decimal places as we wanted. All we need is time, patience, and extreme care in checking all results. Lest anyone should think the process is easy, I suggest the simple exercise of working out the square root of 2 to forty decimal places, with adequate checks on the working.

Anyone who carries the extraction of a single square root to a considerable degree of accuracy may begin to have some appreciation of the enormous labour undertaken by Ludolph van Ceulen, who calculated the value of $\pi$ to 35 places by means of a formula similar to that of Pythagoras. Ludolph lived at Leyden about 1600, and he had his value for $\pi$ engraved on his tomb. Ludolph spent the greater part of a fairly long life in making that calculation. If he had lived a little later, he might have carried the calculation a good deal further.

It was James Gregory, about 1670, who began the process of making the computation of $\pi$ easier. He showed that:

$$\frac{\pi}{4} = 1 - \frac{1}{3} + \frac{1}{5} - \frac{1}{7} + \frac{1}{9} - \frac{1}{13} + \dots.$$

That is a very exasperating series because it converges very slowly, but other series have been developed from it.

Gregory's original theorem was:

$$\tan^{-1} x = x - \frac{1}{3}x^3 + \frac{1}{5}x^5 - \frac{1}{7}x^7 + \dots.$$

That is, the angle whose tangent is $x$ (in radians) is equal to that series. The angle whose tangent is 1 is $45°$ or $\frac{\pi}{4}$, so that if we write $x = 1$, we get the series for $\frac{\pi}{4}$.

Euler used the fact that:

$$\tan (A + B) = \frac{\tan A + \tan B}{1 - \tan A \tan B}.$$

If we make $\tan A = \frac{1}{2}$ and $\tan B = \frac{1}{3}$, then:

$$\frac{\tan A + \tan B}{1 - \tan A \tan B} = \frac{\frac{1}{2} + \frac{1}{3}}{1 - \frac{1}{2} \times \frac{1}{3}} = 1 = \tan \frac{\pi}{4}.$$

So $A + B = \frac{\pi}{4}$, and therefore:

$$\frac{\pi}{4} = \tan^{-1} \frac{1}{2} + \tan^{-1} \frac{1}{3}.$$

We fill in the values of $\tan^{-1}\frac{1}{2}$ and $\tan^{-1}\frac{1}{3}$ from Gregory's series, and so we obtain:

$$\frac{\pi}{4} = \frac{1}{2} - \frac{1}{3}\left(\frac{1}{2}\right)^3 + \frac{1}{5}\left(\frac{1}{2}\right)^5 - \frac{1}{7}\left(\frac{1}{2}\right)^7 + \dots$$
$$+ \frac{1}{3} - \frac{1}{3}\left(\frac{1}{3}\right)^3 + \frac{1}{5}\left(\frac{1}{3}\right)^5 - \frac{1}{7}\left(\frac{1}{3}\right)^7 + \dots.$$

We write down the odd powers of $\frac{1}{2}$, beginning with $\frac{1}{2} = \cdot 5$, and continually dividing by 4; we sort the powers out, alternately $+$ and $-$.

| + | − |
|---|---|
| $\frac{1}{2} = \cdot5$ | $\left(\frac{1}{2}\right)^3 = \cdot125$ |
| $\left(\frac{1}{2}\right)^5 = \cdot03125$ | $\left(\frac{1}{2}\right)^7 = \cdot0078125$ |
| $\left(\frac{1}{2}\right)^9 = \cdot001953125$ | $\left(\frac{1}{2}\right)^{11} = \cdot00048828125$ |

and so on. We carry the division just as far as we want. We then divide by 1, 3, 5, 7, etc.: these numbers are the indices of the powers.

| + | − |
|---|---|
| $\cdot5$ | $\cdot041666666667$ |
| $\cdot00625$ | $\cdot001116071429$ |
| $\cdot000217013889$ | $\cdot000044389205$ |

and so on again. We continue with the powers and divisions till we are clear of the range of accuracy we want.

We deal in the same way with the series of odd powers of $\frac{1}{3}$. We make the necessary additions and subtractions, and finally multiply by 4 to find $\pi$.

Many other variants have been used in order to lighten the labour of calculation. Machin used the identity:

$$\frac{\pi}{4} = 4\tan^{-1}\frac{1}{5} - \tan^{-1}\frac{1}{239}$$

Division by the square of 239 is rather awkward, but that does not matter if we want a comparatively few figures in the result. And there is the considerable advantage of having to find a few terms only. Division by 25, the square of 5, is very simple; we multiply by 4 and move the decimal point two places to the left. Here is the complete calculation:

| + | − |
|---|---|
| $\frac{1}{5} = \cdot2$ | $\left(\frac{1}{5}\right)^3 = \cdot008$ |
| $\left(\frac{1}{5}\right)^5 = \cdot00032$ | $\left(\frac{1}{5}\right)^7 = \cdot0000128$ |
| $\left(\frac{1}{5}\right)^9 = \cdot000000512$ | $\left(\frac{1}{5}\right)^{11} = \cdot0000000 2048$ |
| $\left(\frac{1}{5}\right)^{13} = \cdot00000000 0819$ | $\left(\frac{1}{5}\right)^{15} = \cdot000000000033$ |

After division we have:

| + | − |
|---|---|
| ·2 | ·002666666667 |
| ·000064 | ·00001828571 |
| ·00000056889 | ·000000001862 |
| ·00000000063 | ·000000000002 |
| ·200064056952 | ·002668497102 |
| − ·002668497102 | |
| ·197395559850 | |
| 4 | |
| ·789582239400 | |

$(\frac{1}{239})^3 = ·000000073250$          $\frac{1}{239} = ·004184100418$

$\frac{1}{3}$ of $(\frac{1}{239})^3 = ·00000024417$          $(\frac{1}{239})^5 = ·000000000001$

·789582239400
+ ·00000024417

·789582263817
− ·004184100418

·785398163399
4

3·141592653596

The value of $\pi$ to 12 figures is 3·14159265359.

It would not be a great labour to carry the calculation to the 35 figures that van Ceulen achieved. To carry the working to 40 figures would entail working out each term to 40 places of decimals, and would entail the use of many more terms of the series: 27 terms in the first series and 8 terms in the second series The calculation could be made in a day, as contrasted with the lifetime that van Ceulen devoted to it.

# CHAPTER 13

## *The Ghost Quantity*

As soon as we start dealing with square roots we come up against the root of minus one, $\sqrt{-1}$. Some outrageous nonsense has been talked and written about $\sqrt{-1}$; not quite so much as about $\pi$ and the 'fourth dimension', but bad enough. Mathematically $\sqrt{-1}$ is the quantity which multiplied by itself gives $-1$. Non-mathematically it may be anything from a ghost, or something of spiritual significance, to a meaningless absurdity.

Schoolboys meet $\sqrt{-1}$ in the solution of quadratic equations. Indeed, if we set down a quadratic at random it is as likely as not that its solution will include $\sqrt{-1}$. We learn to call such solutions 'unreal' or 'imaginary', and we do not expect to obtain them from a graph. Nevertheless, the graph of a quadratic does give the imaginary solutions.

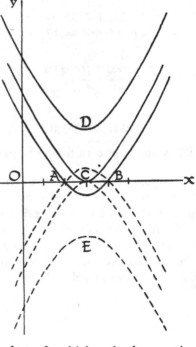

In this diagram there are three whole-line graphs; we ignore the dotted graphs for the moment. The lowest of them is the graph of $x^2 - 6x + 8$; it gives the two values of $x$ that make the equation $x^2 - 6x + 8 = 0$ true. These values are $x = 2$, at $A$, and $x = 4$, at $B$. The next graph above is the graph of $x^2 - 6x + 9$, or $(x-3)^2$. This graph gives the two coincident values of $x$ which make the equation $x^2 - 6x + 9 = 0$ true. These values are 3, at $C$.

The third graph is the graph of $x^2 - 6x + 16$. It gives the solutions of the equation $x^2 - 6x + 16 = 0$. The lowest points of the three graphs are exactly one over another. The point $D$ is 3 units to the right of the zero point, $O$, just as $C$ is. 3 is the real part of the solutions. For the imaginary part of the solutions we find the height of $D$ above the $x$-axis. This height is 2·646, or $\sqrt{7}$. $\sqrt{-7}$ is the imaginary part, so that the full solution is $x = 3 \pm \sqrt{-7}$.

Actually the graph gives the solution $3 + \sqrt{-7}$ only. The solution $3 - \sqrt{-7}$ appears to be missing. That is due to a rather careless way we have of graphing equations. We talk of the 'graph of an equation' when we should say the 'graph of a function'.

$$-x^2 + 6x - 16 = 0$$

is the same equation as:

$$x^2 - 6x + 16 = 0.$$

But $-x^2 + 6x - 16$ is not the same function as $x^2 - 6x + 16$. For a full solution of the equation we have to graph both functions. The lowest of the dotted graphs is the graph of $-x^2 + 6x - 16$. This graph gives the solution $x = 3 - \sqrt{-7}$ at the point $E$.

We may note in passing that the graphs of $-x^2 + 6x - 8$ and $-x^2 + 6x - 9$, the other two dotted graphs, give the same solutions as $x^2 - 6x + 8$ and $x^2 - 6x + 9$. The graphs are different except for the coincident points which give the solutions. The reason is that the equations are the same, and therefore have the same solutions.

The meaning of $\sqrt{-1}$ should now be clear. $+$ is an instruction to measure to the right from the zero point; $-$ is an instruction to measure to the left. $+\sqrt{-1}$ is an instruction to measure up, and $-\sqrt{-1}$ to measure down. $3 + 5\sqrt{-1}$ is a point 3 units to the right and 5 units up. $-4 - 6\sqrt{-1}$ is a point 4 units to the left and 6 units down.

In other words $\sqrt{-1}$ is an operator; just as much an operator as $+$ or $-$ or $\sqrt{\ }$ or $\frac{d}{dx}$. I cannot help thinking that the 1 is an unnecessary intrusion, and that the true symbol is $\sqrt{-}$. The 1 is a mathematical relic that has not been completely disentangled from the symbol; just as we usually write $(-1)^n$ when what we want is the sign only, which should be written $(-)^n$. The symbol

$i$ is often written for $\sqrt{-1}$ or $\sqrt{-}$; when we employ the symbol $i$ we tacitly ignore the 1, except when we want it.

We have come to a conclusion about the significance of $i$ which enables us to represent geometrically quantities which include $i$, *complex quantities* as they are called.

The method of representing complex quantities geometrically was given by Jean Robert Argand in 1806, and afterwards developed by other mathematicians.

A complex quantity may be represented by $x + iy$. $x$ and $y$ are both real quantities. There would be no point in having one or both of them imaginary, because we should not know where we were.

Starting with the zero point, $O$, we measure + quantities to the right, in the direction $OA$. We measure − quantities to the left, in the direction $OA_1$. $+i$ quantities we measure up, in the direction $OB$; and $−i$ quantities down, in the direction $OB_1$.

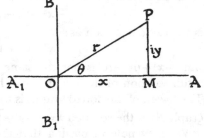

Thus to represent $x + iy$ we measure $x$ from $O$ along $OA$. This takes us to the point $M$. Then we measure $y$ upward from $M$, and so reach $P$. Either the point $P$, or the length $OP$, may be taken to represent the complex quantity $x + iy$. We have done in reverse what we did in finding the imaginary roots of an equation.

If we take $OP$ to represent $x + iy$, we cannot ignore the direction of the line. We usually represent the absolute, or arithmetical length of $OP$ by $r$; $r$ stands for the length $OP$. We use the symbol $\theta$ (theta) for the angle $POM$ which $OP$ makes with $OA$.

By using Pythagoras' Theorem we can find the length of $r$.

$$r^2 = x^2 + y^2,$$
$$r = \sqrt{x^2 + y^2}.$$

Now we use the trigonometrical ratios:

$$\cos \theta = \frac{x}{r}, \quad \text{so } x = r \cos \theta,$$

$$\sin \theta = \frac{y}{r}, \quad \text{so } y = r \sin \theta.$$

We can also remember that $\tan \theta = \frac{y}{x}$, so that $\theta$ is the angle whose tangent is $\frac{y}{x}$. Another way of writing the same thing is $\theta = \tan^{-1} \frac{y}{x}$.

We now have:
$$x + iy = r \cos \theta + ir \sin \theta$$
$$= r (\cos \theta + i \sin \theta).$$

This result turns out to be so useful that we have names for $r$ and $\theta$. $r$ is called the *modulus* of $x + iy$, and $\theta$ is called the *argument*. The modulus is $\sqrt{x^2 + y^2}$ and the argument is $\tan^{-1} \frac{y}{x}$.

Before we can be satisfied with this method of representing complex quantities geometrically we want to see whether it can be used to represent the addition and subtraction of such quantities.

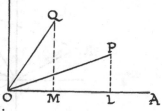

In this diagram $P$ and $Q$ represent complex quantities which we want to add geometrically. In order to distinguish them we call $P$, $x_1 + iy_1$; $Q$ is $x_2 + iy_2$. Then $OL = x_1$, and $PL = y_1$. $OM = x_2$ and $QM = y_2$. $OP$ is $r_1$ and $OQ$ is $r_2$.

The arithmetical sum of $P$ and $Q$ is:
$$x_1 + iy_1 + x_2 + iy_2 = x_1 + x_2 + i (y_1 + y_2)$$

and that is the quantity we want to represent.

We make $LN = OM = x_2$, so that
$$ON = x_1 + x_2.$$

We draw the perpendicular $NR = y_1 + y_2$.

$$R = x_1 + x_2 + i (y_1 + y_2),$$

so that it represents the sum of $P$ and $Q$.

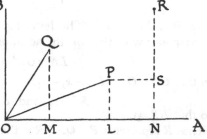

Now let us join $PR$ and $QR$. We thus produce the parallelogram $OQRP$. (We can readily show that triangles $OQM$ and $PRS$ are congruent, and so $PR = OQ$, and is parallel to $OQ$.)

This gives the method of representing $P+Q$. From $P$ we draw PR parallel to $OQ$, and from $Q$ we draw $QR$ parallel to $OP$. $R$, the point of intersection, represents $P+Q$. The modulus of $R$ is:

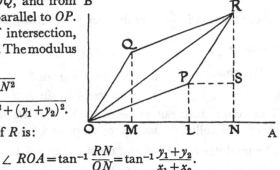

$$OR = \sqrt{RN^2 + ON^2}$$
$$= \sqrt{(x_1 + x_2)^2 + (y_1 + y_2)^2}.$$

The argument of $R$ is:

$$\angle ROA = \tan^{-1}\frac{RN}{ON} = \tan^{-1}\frac{y_1 + y_2}{x_1 + x_2}.$$

To subtract $Q$ from $P$ we have only to draw $PR$ in the opposite

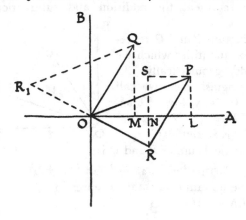

direction, downward. The lettering on the subtraction diagram corresponds with that on the addition diagram.

$$LN = OM = x_2,$$

so that $ON = OL - LN = x_1 - x_2.$

$$RS = y_2 \text{ and } SN = y_1,$$

so that $NR = y_1 - y_2.$

Hence $R$ represents $P - Q$. $OR$ is the modulus, and:

$$OR = \sqrt{ON^2 + NR^2}$$
$$= \sqrt{(x_1 - x_2)^2 + (y_1 - y_2)^2}.$$

The argument is angle $LOR$, and:

$$LOR = \tan^{-1} \frac{NR}{ON} = \frac{y_1 - y_2}{x_1 - x_2}.$$

To subtract $P$ from $Q$ we draw $QR_1$ downward and equal and parallel to $OP$. $R_1$ represents $Q - P$; $OR_1$ is the modulus, and angle $AOR_1$ is the argument. As we should expect, $OR_1$ is equal and opposite to $OR$, since $P - Q = -(Q - P)$.

It is not quite so easy to represent multiplication geometrically. By the usual method of multiplication we have:

$$\begin{array}{r} x_1 + iy_1 \\ x_2 + iy_2 \\ \hline x_1 x_2 + i x_2 y_1 + i x_1 y_2 + i^2 y_1 y_2 \end{array}$$

We know that $i^2 = (\sqrt{-})^2 = -$, so the product is:

$$x_1 x_2 - y_1 y_2 + i\,(x_1 y_2 + x_2 y_1),$$

and that is what we have to represent geometrically.

The method depends on drawing two similar triangles on $OP$ and $OQ$, with one side of the $OP$ triangle equal to one unit.

We make $OL =$ one unit, and join $PL$. For the second triangle we draw a similar triangle on $OQ$, with angle $OQR =$ angle $OLP$. The equal angles are named in the diagram: $a, b, \theta_1$.

Since the triangles are similar we have:

$$\frac{OR}{OQ} = \frac{OP}{OL} = \frac{OP}{1}.$$

$$\therefore OR = OP \cdot OQ.$$

(The idea of making $OL =$ one unit was probably arrived at backwards. $OR \cdot PL = OP \cdot OQ$; so if we make $PL = 1$ we have the result we want.)

That is, $R$ represents the product of $P$ and $Q$. The argument of $R$ is angle $ROA =$ angle $ROQ +$ angle $QOA = \theta_1 + \theta_2$. The modulus of $R$ is:

$$OP \cdot OQ = \sqrt{x_1^2 + y_1^2}\,\sqrt{x_2^2 + y_2^2}$$
$$= \sqrt{(x_1^2 + y_1^2)\,(x_2^2 + y_2^2)}.$$

To represent division, $\dfrac{P}{Q}$, is now comparatively simple; we work in reverse. We make $OL=$ one unit, and join $LQ$. Then we make triangle $OPR$ similar to $OQL$, with angle $OPR=$ angle $OQL$. The equal angles are marked: $a$, $b$, $\theta_2$.

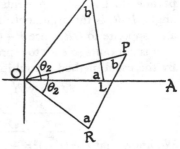

We now have:

$$\frac{OQ}{OL}=\frac{OP}{OR} \quad \text{or} \quad \frac{OQ}{1}=\frac{OP}{OR}.$$

$$\therefore \ OR.OQ=OP$$

and $$OR=\frac{OP}{OQ}.$$

That is, $R$ represents $\dfrac{P}{Q}$. The argument is angle $LOR=$ angle $POR-$ angle $POL=\theta_2-\theta_1$. The modulus is

$$\frac{\sqrt{x_1{}^2+y_1{}^2}}{\sqrt{x_2{}^2+y_2{}^2}}.$$

We will now examine one of the most interesting and exciting theorems connected with complex quantities. The steps in the reasoning are simple enough if we follow them carefully.

We know that:
$$x_1+iy_1=r_1\,(\cos\theta_1+i\sin\theta_1)$$
and $$x_2+iy_2=r_2\,(\cos\theta_2+i\sin\theta_2).$$
So $$(x_1+iy_1)\,(x_2+iy_2)=r_1r_2\,(\cos\theta_1+i\sin\theta_1)\,(\cos\theta_2+i\sin\theta_2).$$

Let us work out the product on the right:

$$\cos\theta_1+i\sin\theta_1$$
$$\cos\theta_2+i\sin\theta_2$$
$$\overline{\cos.\theta_1\cos\theta_2+i\sin\theta_1\cos\theta_2+i\sin\theta_2\cos\theta_1+i^2\sin\theta_1\sin\theta_2}$$
$$=\cos\theta_1\cos\theta_2-\sin\theta_1\sin\theta_2+i\,(\sin\theta_1\cos\theta_2+\cos\theta_1\sin\theta_2).$$

The first part is the well-known equivalent of $\cos(\theta_1+\theta_2)$ and the second part is $i\sin(\theta_1+\theta_2)$.

We now have:
$$(x_1+iy_1)\,(x_2+iy_2)=r_1r_2\,\{\cos(\theta_1+\theta_2)+i\sin(\theta_1+\theta_2)\}.$$

We can treat $r_1r_2$ as if it were a single term, and $\theta_1+\theta_2$ as if it were a single angle; and so we can add a third factor.

$$(x_1+iy_1)\,(x_2+iy_2)\,(x_3+iy_3)$$
$$=r_1r_2r_3\,\{\cos(\theta_1+\theta_2+\theta_3)+i\sin(\theta_1+\theta_2+\theta_3)\}.$$

That is, in the product we multiply the moduli, and add the arguments.

We can proceed step by step until we have multiplied $n$ factors.

$$(x_1 + iy_1)(x_2 + iy_2) \ldots (x_n + iy_n)$$
$$= r_1 r_2 \ldots r_n \{\cos(\theta_1 + \theta_2 \ldots + \theta_n) + i \sin(\theta_1 + \theta_2 \ldots + \theta_n)\}.$$

The theorem begins to get exciting when we make all the factors alike.

$$x_1 = x_2 = \ldots = x_n = x, \quad y_1 = y_2 = \ldots = y_n = y,$$

and so $r_1 = r_2 = \ldots = r_n = r$, and $\theta_1 = \theta_2 = \ldots = \theta_n = \theta$.

When we fill in these values we have on the left $(x + iy)^n$, which is the same thing as $r^n (\cos \theta + i \sin \theta)^n$. On the right we have $r^n (\cos n\theta + i \sin n\theta)$. And so:

$$r^n (\cos \theta + i \sin \theta)^n = r^n (\cos n\theta + i \sin n\theta).$$

Cancelling $r^n$, we have:

$$(\cos \theta + i \sin \theta)^n = \cos n\theta + i \sin n\theta.$$

That is the famous theorem of de Moivre. We have shown that it is true when $n$ is a positive whole number. It can also be shown that it is true when $n$ is a fraction or negative. So it is universally true.

De Moivre's theorem was the key to a whole new world of imaginary or complex trigonometry. The word 'imaginary' is used in the mathematical sense; there is nothing unreal or imaginary in the ordinary sense, nothing that cannot be represented geometrically. The results are as straightforward as other mathematical results.

There is an easy way of sorting out complex results, so that in the end we can have all our results 'real'. Suppose we have:

$$a + ib = c + id,$$

that is, we have two complex quantities equal.

Then
$$a - c = i(d - b).$$

We have a real quantity equal to an imaginary quantity; a measurement in the direction $OA$ is the same thing as a measurement in the direction $OB$ at right angles. That can only be true if both sides of the equation are zero.

$$a - c = 0, \quad d - b = 0;$$

so
$$a = c, \quad b = d.$$

That is, if we have two complex quantities equal, the real parts are equal, and the imaginary parts are equal.

When we are multiplying complex quantities we often have powers of $i$ included in the product.

$$i=i, \quad i^2=-, \quad i^3=-i, \quad i^4=(-)^2=+, \quad i^5=+i \text{ or } i, \text{ and so on.}$$

So we have $i$, $-$, $-i$, $+$ in rotation.

Let us see how this works out.

We know that:

$$\cos n\theta + i \sin n\theta = (\cos \theta + i \sin \theta)^n.$$

We can expand the expression on the right by the binomial theorem, which says that:

$$(x+y)^n = x^n + nx^{n-1}y + \frac{n(n-1)}{\lfloor 2} x^{n-2}y^2 + \frac{n(n-1)(n-2)}{\lfloor 3} x^{n-3}y^3 + \dots.$$

So

$$(\cos \theta + i \sin \theta)^n$$

$$= \cos^n \theta + in \cos^{n-1}\theta \sin \theta - \frac{n(n-1)}{\lfloor 2} \cos^{n-2}\theta \sin^2\theta$$

$$- i \cdot \frac{n(n-1)(n-2)}{\lfloor 3} \cos^{n-3}\theta \sin^3\theta + \dots.$$

We know from de Moivre's theorem that this series is equal to $\cos n\theta + i \sin n\theta$. We sort out the real and imaginary parts, and so we find:

$$\cos n\theta = \cos^n \theta - \frac{n(n-1)}{\lfloor 2} \cos^{n-2}\theta \sin^2\theta$$

$$+ \frac{n(n-1)(n-2)(n-3)}{\lfloor 4} \cos^{n-4}\theta \sin^4\theta - \dots,$$

$$\sin n\theta = n \cos^{n-1}\theta \sin \theta - \frac{n(n-1)(n-2)}{\lfloor 3} \cos^{n-3}\theta \sin^3\theta$$

$$+ \frac{n(n-1)(n-2)(n-3)(n-4)}{\lfloor 5} \cos^{n-5}\theta \sin^5\theta - \dots.$$

These are master equations that give the values of $\cos 2\theta$, $\sin 2\theta$, $\cos 3\theta$, $\sin 3\theta$, etc. in terms of the sines and cosines of $\theta$, and their powers. We have only to put $n=2$, 3, 4, etc. in the equations.

$$\cos 2\theta = \cos^2 \theta - \frac{2 \times 1}{\lfloor 2} \cos^0 \theta \sin^2\theta.$$

All the other terms include the factor $n-2=0$, and so they vanish.

The zero power of any quantity $=1$; and so $\cos^0\theta=1$. The equation becomes:

$$\cos 2\theta = \cos^2\theta - \sin^2\theta = 2\cos^2\theta - 1.$$

For $\sin 3\theta$ we put $n=3$.

$$\sin 3\theta = 3\cos^2\theta \sin\theta - \frac{3\times 2\times 1}{\underline{|3}}\cos^0\theta \sin^3\theta.$$

(All the other terms include $n-3=0$.)

$$= 3\cos^2\theta \sin\theta - \sin^3\theta.$$

We can substitute for $\cos^2\theta$ its equivalent $1-\sin^2\theta$, and so obtain the value of $\sin 3\theta$ in terms of $\sin\theta$ and its powers:

$$\sin 3\theta = 3(1-\sin^2\theta)\sin\theta - \sin^3\theta$$
$$= 3\sin\theta - 4\sin^3\theta.$$

There is a very interesting and rather extraordinary way in which we can transform the two equations. Instead of $n\theta$ we write $x$. The point is that we can make $n$ as big as we like, and then by making $\theta$ correspondingly small we can have $x$ as a finite number.

$$n\theta = x, \quad \text{so } n = \frac{x}{\theta}.$$

With these changes the equation for $\sin n\theta$ becomes:

$$\sin x = \frac{x}{\theta}\cos^{n-1}\theta \sin\theta - \frac{\dfrac{x}{\theta}\left(\dfrac{x}{\theta}-1\right)\left(\dfrac{x}{\theta}-2\right)}{\underline{|3}}\cos^{n-3}\theta \sin^3\theta + \ldots.$$

(We move the $\theta$'s so as to be under $\sin\theta$ and its powers.)

$$= x\cos^{n-1}\theta\,\frac{\sin\theta}{\theta} - \frac{x(x-\theta)(x-2\theta)}{\underline{|3}}\cos^{n-3}\theta\,\frac{\sin^3\theta}{\theta^3} + \ldots.$$

We are going to make $\theta$ very small, so $\sin\theta = \theta$, the circular measure of the angle; and $\dfrac{\sin\theta}{\theta}=1$. We can replace by $1$, $\dfrac{\sin\theta}{\theta}$ and all its powers.

$$\sin x = x\cos^{n-1}\theta - \frac{x(x-\theta)(x-2\theta)}{\underline{|3}}\cos^{n-3}\theta + \ldots.$$

That is not all that happens. When $\theta$ is negligibly small, $\cos\theta = 1$, and so do all its powers. And we can ignore $\theta$, $2\theta$, etc. in the factors.

$$\sin x = x - \frac{x^3}{\underline{|3}} + \frac{x^5}{\underline{|5}} - \frac{x^7}{\underline{|7}} + \ldots.$$

We can deal in the same way with the series for cos $n\theta$.

$$\cos x = \cos^n \theta - \frac{\frac{x}{\theta}\left(\frac{x}{\theta} - 1\right)}{\lfloor 2} \cos^{n-2}\theta \sin^2\theta + \dots$$

$$= \cos^n \theta - \frac{x\,(x-\theta)}{\lfloor 2} \cos^{n-2}\theta \frac{\sin^2\theta}{\theta^2} + \dots$$

$$= 1 - \frac{x^2}{\lfloor 2} + \frac{x^4}{\lfloor 4} - \frac{x^6}{\lfloor 6} + \text{etc.}$$

We thus get important and valuable series for sin $x$ and cos $x$. We can use them in calculating tables of the trigonometrical functions, or when we want the sine or cosine of an angle not in the tables. We begin by changing the angle whose sine or cosine we want to find into radians.

$$180° = \pi \text{ radians} = 3\cdot14159 \text{ radians},$$

$$1° = \frac{3\cdot14159}{180} = \cdot017453 \text{ radian},$$

$$\sin 1° = \left(x - \frac{x^3}{\lfloor 3} + \dots\right)$$

$$= \cdot017453 - \frac{\cdot017453^3}{\lfloor 3} +$$

$$= \cdot017453 - \cdot0000009$$

$$= \cdot017452.$$

I hope, rather against hope, that I have succeeded in laying the ghost of $\sqrt{-1}$, and in showing that there is nothing unreal or imaginary about it in the ordinary senses of the words. It is an ordinary mathematical symbol, capable, as we have seen, of geometrical representation. It can be measured in the direction where it exists, and not in the direction where it does not exist. Just as ordinary arithmetical quantities can be measured in the direction in which they exist, but not in the direction in which they have no existence.

Mathematicians aim at being simple, clear, open. They set out their results plainly so that all who take sufficient trouble of the right kind may read and understand. But there seems to be no limit to the perversity of people who seek to darken counsel by words without wisdom, and especially of those who seek to pervert the plain truths of astronomy and mathematics.

# CHAPTER 14

## *Poets' Numbers*

POETS ARE not always quite at home with numbers. And that is rather odd, because poetry is mathematical prose. There may be other differences between prose and poetry; but I am not convinced that the high emotional content of poetry is un-connected with the fine mathematical precision that is demanded of poets.

Poets have to be able to count in threes; unless, that is, they confine their writings to iambs and trochees. Mary Howitt was even able to count in fours when she wrote:

'Will you walk into my parlour?' said a spider to a fly.

Though her counting did get rather sketchy when she came to the second line:

''Tis the prettiest little parlour that ever you did spy.'

'Prettiest' seems to be used as a two-syllable word, and 'parlour' as three syllables; not the other way about as one might have expected. After that, she probably gave up 'parlour' as a bad job; though some of the later lines do recapture the fine careful rapture of the first.

Lewis Carroll must have admired Mary Howitt's unusual, if sketchy, ability to count in fours. He imitated the careful counting of the first line in:

'Will you walk a little faster?' said a whiting to a snail.

Then he saw the snag that had wrecked her counting; he avoided it by steering an entirely new course. As one would expect from a mathematician, he got his counting correct in the second line:

'There's a porpoise close behind us and he's treading on my tail.'

It is not till he comes to the third verse that Lewis Carroll begins to weary of the desperate effort of counting in fours. He docks two unaccented syllables from the beginnings of the lines, instead of one as in the other verses. In the third line he is right up against the 'parlour' difficulty; and he succumbs.

> The further off from England the nearer is to France.

'England' is the same sort of word as 'parlour', and he makes it do as a three-syllable word.

Poets have indeed chosen hard ways. It is not easy to count in fours, or even in threes or twos, especially when one has to consider the colouring effect of long and short vowels. It is no wonder that Hamlet wrote to his dear Ophelia, 'I am ill at these numbers'. He might well be.

It is not merely that poets have to count in twos or threes or fours; indeed great poets do that by instinct. Like Pope, they lisped in numbers, for the numbers came. Though Browning found it necessary in later life to 'mend the metre a little' of his first published work.

There are patterns in poetry that are essentially arithmetical. The sonnet, with its octet and sestet, or some other arrangement, is a form voluntarily accepted by the poet who writes it, and faithfully adhered to. The arithmetical pattern, firmly held in the mind, becomes not a master but a servant. The writers of free verse have sacked the arithmetical servant, dreading that the servant might become master. And that looks like a confession that the free verse writers do not claim to be amongst the 'great masters'.

When actual numbers come into poetry they can be used well or ill. Some poets use numbers very seldom. Keats' numbers hardly seem to matter. The only ones that stand out are:

> And there I shut her wild wild eyes
> > With kisses four.

and:

> Oh joy! for now I see a thousand eyes
> Wide glaring for revenge.

and perhaps the 'dazzled thousands' of *Endymion*.

Keats' few numbers are nearly always a thousand for a large number, and a million for a very large number. His sky is over-spangled with a million of little eyes; though a thousand would be nearer the truth. The spider's shuttle circled a million times within the space of a swallow's nest-door; and that is, I fear, an exaggeration that casts suspicion on all Keats' millions.

Wordsworth uses numbers with certainty. His 'platform eight feet square' is exactly that size. Just as·we know that Simon Lee's scrap of land is 'not twenty paces from the door' of their moss-grown hut of clay, or that 'six feet in earth my Emma lay'. His times are numerical, with a little added touch of the poet's grace, but accurately numerical: 'the suns of twenty summers danced along', 'five summers with the length of five long winters', 'nine tedious years'. Even when he is being extravagant Wordsworth is not unduly extravagant. The cuckoo made him look a thousand ways; he heard a thousand blended notes while in a grove he sat reclined; the sexton's bonehouse contained three thousand skulls. These numbers are as real as the thousand years the Strid hath borne that name.

Shakespeare, as one would expect of him, took numbers in his stride; he put them in as they came. Henry V reads the tale of the French slain at Agincourt:

> This note doth tell me of ten thousand French
> That in the field lie slain: of princes in this number,
> And nobles bearing banners, there lie dead
> One hundred twenty-six; added to these
> Of knights, esquires, and gallant gentlemen,
> Eight thousand and four hundred; of the which
> Five hundred were but yesterday dubb'd knights;
> So that in these ten thousand they have lost
> There are but sixteen hundred mercenaries.

The arithmetic is hardly up to modern standards. Here it is:

$$\begin{array}{rl}
10000 & \text{that in the field lie slain} \\
\text{minus} \quad 126 & \text{princes, and nobles bearing banners} \\
\text{minus} \quad 8400 & \text{knights, esquires, and gallant gentle-} \\
& \text{men}
\end{array}$$

leaving unexpectedly 1600 mercenaries.

The bitter arithmetical fool in *King Lear* promises:

> Thou shalt have more
> Than two tens to a score.

Perhaps Henry's was that kind of arithmetic. Or perhaps he was confused by the irrelevant intrusion of 500 new-dubbed knights into the sum. That is surprising when we consider Henry's nice appreciation of inverse proportion:

> The fewer men the greater share of honour.

Kipling uses numbers with certainty, and he does get his sums right.

> Seven—six—eleven—five—nine-an'-twenty mile to-day—
> Four—eleven—seventeen—thirty-two the day before—
> (Boots—boots—boots—boots, movin' up and down again!)

The arithmetic is of course:

$$7+6+11+5=29, \text{ and } 4+11+17=32.$$

Numbers are essential parts of Kipling's poems, and he often repeats them, sometimes as a refrain. 'We've got the cholera in camp' runs to the refrain of 'ten deaths a day'. In 'The Young Queen', who is Australia, he repeats the numbers five and a thousand, and he is intrigued with the idea of:

> The Five-starred Cross above them, for sign of the Nations Five.
> So it was done in the Presence—in the Hall of our Thousand Years,
> In the Face of the Five Free Nations that have no peer but their
>      peers.

The place of Bunyan's *Holy War* he fixes with a number and a date:

> Two hundred years and thirty
> Ere Armageddon came.
>
> .    .    .    .    .    .
>
> The craft that we call modern,
>      The crimes that we call new,
> John Bunyan had 'em typed and filed
>      In Sixteen Eighty-two.

'The Man of Sixty Spears' is a sufficient description of the Chief in War who held the Pass that leads to Thibet. And the Priest of Kysh was 'bent with a hundred winters'.

> Who remembers
> Forty-odd-year old Septembers?

is an arithmetical jest in Browning's manner.

And so on, to Kipling's arithmetical 'Lesson':

We have spent two hundred million pounds to prove the fact
    once more,
That horses are quicker than men afoot, since two and two
    make four.

Tennyson could use numbers dramatically.

Into the mouth of hell
Rode the six hundred!

has caught the thunder of the hooves. 'Scarlett and Scarlett's
three hundred' was less successful; it was a mistake to try repetition
when plain repetition was plainly ruled out.

Spanish ships of war at sea,
We have sighted fifty-three!

gains its effect not only from the size but from the definiteness of
the number. Fifty or sixty might have been mere exaggerations;
but there is no getting away from fifty-three. Those ships were
counted: one, two, three, up to the amazing total.

The 'Ballad of the Fleet' goes on insisting on numbers: 'her
hundred fighters on deck, and her ninety sick below', 'but never
a moment ceased the fight of the one and the fifty-three'. Only
in the night-long battle do the definite numbers give place to the
reiterated 'ship after ship the whole night long', and that is
because each galleon was an individual menace.

There are numbers that seem poetically right. There is a quality
about seven. Six Richmonds in the field were one too many; but
usually seven is the poet's number. It is getting near the limit of
easy visualisation. Eight is two fours, nine is three threes, and
ten is two fives. But seven we have to accept as seven, without the
prosaic aid of factorisation.

The seven days of the week have each an individual character,
ranging from the quiet of Sunday to the riot of Saturday. If the
week were eight days, we should at once begin to halve and
quarter it; but seven is seven individual days.

There were seven wonders of the world; the eighth wonder, so
often added, is nothing but a nine-days wonder. There were
seven heavens; an eighth heaven would never have been used
'to flatter beauty's ignorant ear'. There were seven sleepers of

Ephesus; well, let them sleep. Seven heroes went up against
Thebes, and there were seven champions of Christendom. The
name Seven Dials remains, although the clock with seven dials,
facing seven converging streets, is gone. There had to be a trial
of seven bishops, seven sages, seven liberal arts, a Seven Years'
War, and every other seven that could be thought of, including
seven-leagued boots. The seventh wave was probably the imagination of some ancient mariner. But the 'seven seas from marge to
marge' are the seven seas of the practical navigator: the North
and South Atlantic, the North and South Pacific, the Indian Sea,
the China Sea, and the Caribbean.

When Peter came to Jesus to ask how often he should forgive
his brother, he himself suggested the poet's number; and he
evidently meant a magnificent range of forgiveness. But Jesus
said unto him, I say not unto thee, Until seven times: but, Until
seventy times seven. And that puts no limit to forgiveness.

Swinburne stole the numbers, and used them to deck his
Dolores:

> Seven sorrows the priests give their Virgin;
>     But thy sins which are seventy times seven,
> Seven ages would fail thee to purge in,
>     And then they would haunt thee in heaven.

When Rossetti's Blessed Damozel leaned out from the gold bar
of Heaven:

> She had three lilies in her hand
> And the stars in her hair were seven.

The Ancient Mariner saw the curse in dead men's eyes.

> Seven days, seven nights, I saw that curse
> And yet I could not die.

And later on:

> Like one that hath been seven days drowned
> My body lay afloat.

But Coleridge's sevens may be merely weeks.

Next to seven comes three. It is the smallest number that
ensures a majority on one side or the other on all occasions. That
is why triumvirs were always popular; though one can hardly

imagine the slight unmeritable Lepidus holding the balance between Antony and Octavius. It is no wonder Antony exclaims:

> Is it fit,
> The three-fold world divided, he should stand
> One of the three to share it?

Lear had to have three daughters. Two daughters might have been just the clash of one personality against another. If there had been four they would have begun to go in twos, like the legs of a table on an uneven floor. No, there had to be three. Just as Two-eyes had to stand up against One-eye and Three-eyes. There has to be a safe majority for evil, in order that virtue may at first be vanquished; and three is the number that ensures that.

The number three runs all through *The Merchant of Venice*. Antonio borrowed three thousand ducats, and for three months. He expected thrice three times the value of the loan, and Bassanio offered Shylock thrice his principal. The thrice-fair Portia had three suitors, and three of Antonio's argosies came finally to port.

Macaulay's dauntless three held the bridge in the brave days of old; and it was the worth of three kingdoms Cowper would defy, to lure him to the baseness of a lie. In Hardy's 'Leipzig':

> Around the town three battles beat
> Contracting like a gin.

James Thomson's *City of Dreadful Night* had but three terms, dead Faith, dead Love, dead Hope. Milton barred the gates of Hell with gates thrice threefold.

> Three folds were Brass,
> Three Iron, three of Adamantine Rock,
> Impenetrable, impal'd with circling fire,
> Yet unconsumed.

The three imperial crowns in Browning's *Sordello* were:

> Aix' iron, Milan's silver, and Rome's gold.

And last comes the ringing challenge at the end of *King John*:

> Now these her princes are come home again,
> Come the three corners of the world in arms
> And we will shock 'em.

Though five has had less fortune as a poet's number, nevertheless Rossetti's Lady Mary had five handmaidens:

> whose names
> Are five sweet symphonies:—
> Cecily, Gertrude, Magdalen,
> Margaret and Rosalys.

Eleven has its place in the inaccurate expression 'the eleventh hour', Kipling's 'dark eleventh hour'. The hour intended is the twelfth hour, the last before midnight.

'Twenty' is little used by poets. 'Score' is preferred, perhaps because it has more rhymes, and perhaps because it is a single syllable and more readily slipped into a line of verse. But is there a prettier line anywhere than:

> Then come kiss me, sweet and twenty?

Even though 'plenty' was rather forced into the previous line, 'In delay there lies no plenty', to make way for it.

Nine is a resonant number. Lars Porsena of Clusium by the Nine Gods he swore; and to leave no doubt about it Macaulay repeated 'by the Nine Gods he swore it'. It is rather surprising that the number nine has been used so sparingly; but it has made spectacular appearances, usually as part of a number. Browning's Last Duchess:

> Thanked men,—good! but thanked
> Somehow—I know not how—as if she ranked
> My gift of a nine-hundred-years-old name
> With anybody's gift.

How magnificently the number is spread out!

In praise of his own county of Sussex, Kipling with equal soul would see:

> Her nine-and-thirty sisters fair,
> Yet none more fair than she.

I once had to edit an article in which Kipling's nine-and-sixty ways of constructing tribal lays was misquoted as four-and-twenty. Four-and-twenty has nothing to do with tribal lays; but it is right for blackbirds baked in a pie; it is the exact number that a generous-minded cook would put in it.

John Buchan chose well in the title of his *Thirty-nine Steps*.

William Blake was dubious about Chaucer's count of his own pilgrims:

> Full nine and twenty in a company.

His own careful count made it thirty-one, including Chaucer himself. He says, The Webbe, or Weaver, and the Tapiser, or Tapestry Weaver, appear to me to be the same person; but this is only an opinion, for "full nine and twenty" may signify one more or less. But I daresay that Chaucer wrote 'A Webbe Dyer', that is a Cloth Dyer:

> A Webbe Dyer and a Tapiser.

Perhaps the most spectacular appearance of nine is in Browning's:

> There's a great text in Galatians
> Once you trip on it entails
> Twenty-nine distinct damnations,
> One sure if another fails.

The great text begins with the 19th verse of the fifth chapter of *Galatians*. I never could make it add up to twenty-nine. Seventeen works of the flesh are enumerated, ending with 'and such like'. In addition there are nine fruits of the Spirit against which one might offend. $17 + 9 = 26$. Browning took such extreme pains to be accurate that, in spite of the count, I felt sure he must be right. I went back humbly to make a final count, and found I had stopped counting too soon. There are three more.

But Browning's number is fifty. When he is not tied by fact to a particular number, and can let his fancy roam, fifty is the number he chooses. In a poem addressed to his wife he writes:

> There they are, my fifty men and women
> Naming me the fifty poems finished.

And in 'A Likeness' he says:

> I keep my prints, an imbroglio,
> Fifty in one portfolio.

When the Pied Piper asked for a thousand guilders:

> 'One? Fifty thousand!' was the exclamation
> Of the astonished Mayor and Corporation.

And when they tried to beat him down:

> 'A thousand guilders! Come, take fifty!'

In *The Ring and the Book* Pompilia's seventeen years are all-important; but by way of contrast there are men of fifty or thereabouts. Guido has his exact official age:

> Worn threadbare of soul
> By forty-six years rubbing on hard life.

But Violante has 'flagrant fifty years'; though later on her age is given as:

> Far-over-fifty years
> Matching his sixty-and-under.

It is a man of fifty who finds his corns ache and his joints throb, and foresees a storm. As for men in general:

> They get to fifty and over: how's the lamp?
> Full to the depth of the wick.

And having got through 'fifty years of flare', they burn out so.

At her marriage Pompilia stood only with fifty candles company. When a man is feeing friends for a murder, Archangelis suggests 'Why not take fifty?' Bottinius would parade his studies fifty in a row. The Pope declares that Judas could not plead:

> He was too near the preacher's mouth, nor sat
> Attent with fifties in a company.

Guido exclaims bitterly:

> Go bid a second blockhead like myself
> Spend fifty years in guarding bubbles of breath.

With equal bitterness he cries out against 'your muster of some fifty monks', 'fifty times the number squeak and gibber in the madhouse', and against fifty miracle-mongers. And lastly:

> You have some fifty servants, Cardinal,—
> Which of them loves you?

In *Sordello* Richard agrees to enter Ferrara, flanked with fifty chosen friends. Eclin had to share himself:

> In fifty portions, like an o'ertasked elf
> That's forced illume in fifty points the vast
> Rare vapour he's environed by.

In the exhortation 'Sordello awake!' there is the cry:

> For one thrust forward, fifty such fall back.

And Berta, on that chance heap of wet filth, is reconciled to fifty gazers.

Somehow I have a feeling that Browning was always fifty years old.

Milton's numbers are nearly always powers of ten. The shape of Death is fierce as ten Furies. The grieslie terrour at Hell's gate grew ten-fold more dreadful and deform. The leaders of the Rebel Angels with hundreds and with thousands came attended to the conclave in Pandæmonium, whereat a thousand Demy-Gods sat on golden seats. Milton thinks he might relate of thousands who fought with Gabriel; and Gabriel led forth to Battel the armed Saints by Thousands and by Millions rang'd for fight. Eve had a thousand decencies.

Satan, questing through the wilde Abyss:

> All unaware
> Fluttering his pennons vain plumb down he drops
> Ten thousand fadoms deep.

Ten thousand Thunders are hurled against the rebels. Death Ministers, not Men, multiply ten thousand fould the sin of him who slew his Brother.

> Heav'n Gates
> Poured out by millions her victorious Bands
> Pursuing.

Millions of fierce encountering Angels fought on either side. The moon:

> Her Reign
> With thousand lesser Lights dividual holds,
> With thousand thousand Starres, that then appear'd
> Spangling the Hemisphere.

The numbers used by poets are moderate enough; they seldom exaggerate. The tremendous theme of *Paradise Lost* seems to call for big numbers, but Milton is sparing of them. He rather chooses similes like the Autumnal Leaves that strow the Brooks of Vallombrosa, or scattered sedge afloat. The biggest of his definite numbers is ten millions, the 'ten thousand thousand Saints'.

Of big numbers used by poets the most common is ten thousand, or its Greek equivalent 'myriad'. Ten thousand is large but comprehensible.

Richard III says, gloomy but still valiant:

> Shadows to-night
> Have struck more terror to the soul of Richard
> Than can the substance of ten thousand soldiers
> Arm'd in proof.

Juliet cries out in agony:

> That 'banished', that one word 'banished'
> Hath slain ten thousand Tybalds.

Henry V rebukes his cousin Westmoreland for wishing:

> Oh that we now had here
> But one ten thousand of those men in England
> That do no work to-day.

Hamlet doubles the number:

> I see
> The imminent deaths of twenty thousand men
> That for a fantasy or trick of fame
> Go to their graves like beds.

And Clifford, dying in the cause of Henry VI, says:

> I, and ten thousand in this luckless realm,
> Had left no mourning widows for our deaths.

The ten thousand daffodils that Wordsworth, or his sister, saw at a glance, are quite a possibility.

Cowper sees in the garden ten thousand dangers that lie in wait to thwart the process of growth. His enumeration however falls far short of that number:

> Heat, and cold, and wind, and steam,
> Moisture and drought, mice, worms, and swarming flies.

He seems to err on the side of understatement when he writes of the Chief, demideified and fumed, who deems:

> A thousand, or ten thousand lives,
> Spent in the purchase of renown for him,
> An easy reckoning.

And finally he reaches near accuracy with:

> Ten thousand sit
> Patiently present at a sacred song,
> Commemoration-mad.

Million is not very greatly used by poets, except perhaps to describe inaccurately the star-spangled heavens. When Kipling wants to be purposely extravagant he gives us:

> Forty million reasons for failure, but not a single excuse.

And even:

> Strings o' forty thousand million
> Boots—boots—boots—boots, movin' up an' down again.

But such extravagance is unusual.

Shakespeare's most notable reference to a million is probably:

> Since a crooked figure may
> Attest in little place a million.

Tennyson's 'wild mob's million feet' is as accurate as it is vigorous. Browning imagines millions of the wild made tame on a sudden at Waring's fame; Hardy, the million feet of columns closing in at Leipzig; and Coleridge a thousand thousand slimy things.

The ten thousand times ten thousand of the *Revelation* is not so big a number as it sounds. It is a hundred millions, or a twentieth part of the present population of the world. Perhaps the writer was conscious of this, for he adds to it 'and thousands of thousands'.

Poets are rather apt to set us little sums, usually very easy ones. Southey's 'twice four hundred men' who went down with the *Royal George* is one of the easiest; so is Keats' 'thrice five miles of fertile ground'. Kipling's 'strength of twice three thousand horse' is not much harder, nor is his 'thrice thirty thousand, to force the Jumna ford'. Gloster, in *Henry VI*, sets a harder sum:

> And had I twenty times as many foes
> And each of them had twenty times their power.

That would multiply the power of the foes by $20 \times 20 = 400$.

Browning's Kentish Sir Byng who stood for the King, 'marched them along, fifty-score strong'. He might have said a thousand, but that would have robbed him of his recurring fifty.

The hardest sum I can remember in any poem is, oddly enough, in Keats' *Endymion*: 'one million times ocean must ebb and flow'. Allowing a little under twelve hours for a tide this works out at about 1350 years. Keats himself supplies ten centuries as the answer; so he is not so far out as one might have feared.

Wordsworth does work out his sums, and he is especially fond of subtraction. In 'The Last of the Flock' he works up to 'full fifty comely sheep', and then sadly down to 'the last of all my flock'. Though he had no luck with the simple child to whom he tried to teach his own art of subtraction.

When they are well used, numbers can provide the vivid contrasts that poets sometimes seek. Tennyson's 'millions of rifle bullets, and thousands of cannon balls' is disappointing; we expect an upward climb from millions, and we get a climb down instead. He was happier with his contrasts:

Better fifty years of Europe than a cycle of Cathay.

and the triumphant conclusion of 'The Defence of Lucknow':

'Hold it for fifteen days!' we have held it for eighty-seven!

Byron also had a vivid contrast when he wrote:

Ships in thousands lay below
And men in nations, all were his.
He counted them at break of day;
And when the sun set where were they?

Though I am sceptical about the count.

Kipling, who really liked numbers, wrote of the old regime in Russia:

Three hundred years it flourished—in three hundred days it died.

And he wrote of Egypt:

But the Sergeant he had hardened Pharaoh's heart.
That was broke, along of all the plagues of Egypt,
Three thousand years before the Sergeant came—
And he mended it again in a little more than ten
So Pharaoh fought like Sergeant Whatisname!

Shakespeare, who also liked numbers, gives us perhaps the most striking numerical contrast, in *Richard II*:

> Yet I well remember
> The favours of these men: were they not mine?
> Did they not sometimes cry, All hail! to me?
> So Judas did to Christ. But he, in twelve,
> Found truth in all, but one; I, in twelve thousand, none.

Fractions, apart from halves and quarters, and occasional thirds, are almost entirely absent from poetry. Hardy wrote vigorously:

> Give me a third of their strength, I'd fill half Hell with
> their soldiers.

Petruchio, rounding on the Tailor, with his own measures, calls him:

> Thou yard, three-quarters, half-yard, quarter, nail!

But no poet ever set out with a more forlorn hope than Milton when he tried to put into poetry the $23\frac{1}{2}°$ inclination of the earth's axis:

> Some say he bid his Angels turne ascanse
> The Poles of Earth twice ten degrees and more.

Could anyone do better—or worse?

And here is Thomas Hardy turning a number into romance:

> When I set out to Lyonesse,
> A hundred miles away,
> The rime was on the spray,
> And starlight lit my lonesomeness
> When I set out for Lyonesse
> A hundred miles away.

# CHAPTER 15

## *Numbers that mean too much*

THERE ARE so many odd things about numbers that people are apt to read into them more than is actually there. Every mathematician knows that the relations between numbers are arithmetical, and he reads nothing into them except such relations. There is no telling what the non-mathematical will read into numerical relations.

The great name of Pythagoras is attached to an arithmetical philosophy. The philosophy appears in various forms; one could hardly expect two philosophers, even if they belong to the same school, to reach the same result. $1+2+3+4$ is always ten, whatever philosopher is working on it. But the meanings to be attached to 1, 2, 3, 4, and their sum, depend on the fancy of the particular philosopher.

One is one, complete and unchangeable; one is the reason. Two is divisible, and therefore changeable; two is opinion. Four is $2 \times 2$, the first square number (if we omit 0 and 1); four is the product of two equals; four stands for justice. Then again 2 is the female number; 3 is the male number. So $5=2+3$ is the symbol of marriage. Well, why not? and again why?

$1+2+3+4=10$, so ten becomes an extremely important number. It is the sum of all kinds of things, all depending on what meanings are attached to 1, 2, 3 and 4.

We are all fond of oppositions: here and there, now and then, on and off, now or never, sooner or later; it would be easy to make a considerable list. The ancient philosophers took the matter very seriously, and ten being a number of great potency, they had to have ten fundamental oppositions. A Pythagorean set was: 1. Limited and unlimited. That was a magnificent start. Though we are always trying to bridge the gulf between the two; as when we find a finite sum for an infinite series. 2. Odd and even. That seems a terrible come down; but in the infancy of mathematics it might well seem as important as it does in infants'

schools to-day. 3. One and many. Which reminds me of
Artemus Ward's reason for surrendering: 'There were forty of
them, and only one of me.' 4. Right and left. That is no more
than a small convenient way of describing relative positions.
5. Male and female. Which introduces a great part of biology
into the oppositions. 6. Rest and motion. Here we have the
most elementary idea of mechanics. 7. Straight and crooked.
A poor little opposition. 8. Light and darkness. That can be
construed as enlightenment and ignorance. 9. Good and evil.
Which is the whole problem of ethics. 10. Square and oblong.
That at least makes up the ten oppositions, and there is little else
to be said for it.

Now let us see how this potent number, ten, was applied to the
universe. At the centre of the universe there is a great fire, not
as you might think, the sun, but the throne of Zeus. Round this
are ten spheres or heavenly bodies: the sun, the moon, the earth,
the five known planets, the sphere of the fixed stars. That is only
nine, and there had to be ten. So the counter-earth was invented;
this was a dark body which shielded the earth from the fierce
rays of the central bonfire, and allowed us to receive light and
heat only by reflection from the sun, and to a lesser degree from
the moon. The counter-earth is the kind of thing we are driven
to when we try to attach to numbers meanings beyond the
arithmetical warrant.

The five planets known to the Ancients have been increased to
ten: Mercury, Venus, the earth, Mars, the Asteroids (if we count
them as one), Jupiter, Saturn, Uranus, Neptune, Pluto. So we
could go back to that potent number ten, without the ad-
ventitious aid of sun, moon, star sphere, and counter-earth. It
worries me to think of astrologers making out their horoscopes in
ignorance of, or ignoring, quite a lot of 'planetary influences'. It
is no wonder they made the most outrageous errors.

It is so easy to juggle with numbers. We can add them and
subtract them, multiply them and divide them in an infinity of
ways. If a number does not suit our purpose, we can miss it out;
or we can put another number in that will fit. If we are looking
for odd relations or strange coincidences, we can find them by the
bushel. The fact is that we put them there. We get ourselves into

all kinds of tangles and difficulties when we ignore that plain fact, and try to read into such relations and coincidences anything but purely mathematical relations.

Strange that when we multiply 4649 by 239 we should get a number consisting entirely of 1's! Not strange at all; those numbers were discovered by hunting for the factors of 1,111,111.

Strange that when we write 142857, and then move the 1 to the other end, so as to get 428571, we should get a number exactly three times as great. Not strange at all. 142857 are the figures we obtain in changing one-seventh to a decimal; and 428571 are the figures for three-sevenths. We can find just as many relations of that kind as we want.

Strange that this 'magic square' should give the sum 65 in four different ways: row, column, diagonals. But that is not all. The numbers 1, 2, 3, to 25 follow knight's moves exactly over the square. And if we take any two numbers equidistant from the centre, their total is twice the middle number. Thus:

| 23 | 18 | 11 | 6  | 25 |
|----|----|----|----|----|
| 10 | 5  | 24 | 17 | 12 |
| 19 | 22 | 13 | 4  | 7  |
| 14 | 9  | 2  | 21 | 16 |
| 1  | 20 | 15 | 8  | 3  |

$$1 + 25 = 9 + 17 = 19 + 7$$
$$= 10 + 16 = 13 \times 2,$$

and so on for all the other pairs.

With such a mass of coincidences, surely the square has some significance beyond its purely mathematical one. Even simpler squares have been treasured by the credulous for their magical properties. Alas, the name 'magic square' is no more than a jest; and the construction of magic squares is a mild mathematical amusement. They have been constructed by the hundred with all sorts of odd properties. It is quite a simple matter to construct magic squares from which the outside edges of numbers can be removed, leaving a magic square within. Even the remarkable square given above is no more than a *tour de force* by Euler, who himself put into it its odd properties.

If you write down the numbers 1 to 6 in different orders six times, one line under another, it is highly improbable that there will be one and one only of each figure in each row, column and diagonal.

Now look at this square. There are three sets of double figures: 3–5, 1–2, 4–6. If you copy all the single figures, then either the first or the second of the 3–5 pairs, either the first or the second of the 1–2 pairs, and either the first or the second of the 4–6 pairs, then in each case you produce a square with one and one only of each figure in each row, column, and diagonal. The square gives eight such squares, $2 \times 2 \times 2$.

| 1 | 2 | 3 | 4 | 5 | 6 |
|---|---|---|---|---|---|
| 35 | 53 | 6 | 12 | 4 | 21 |
| 46 | 64 | 2 | 53 | 1 | 35 |
| 64 | 46 | 1 | 35 | 2 | 53 |
| 53 | 35 | 4 | 21 | 6 | 12 |
| 2 | 1 | 5 | 6 | 3 | 4 |

There is no value in the square, no special significance attached to it, except for a little amusement in ferreting it out. It is possible to produce fifteen other squares of the same kind as the one given; so that if we start with 1, 2, 3, 4, 5, 6 in the top line we can have $16 \times 8 = 128$ different arrangements which give one and one only of each figure in each row, column and diagonal.

Some of the coincidences arrived at by the circle squarers are not so remarkable as they appear to think. Thus we are given the numbers ·2146 and ·1416. The latter is derived from the former by the process $1 - 4 \times ·2146 = ·1416$. We are told:

$$\frac{1}{·2146} = 4·6598322460\ldots$$

and

$$\frac{·1416}{·2146} = 0·6598322460\ldots.$$

Placing the latter over the former we get:

$$\frac{6598322460}{46598322460} = ·1416.$$

Remarkable—until we analyse the process. For clearness we write $a$ for ·2146; then $·1416 = 1 - 4a$. We have:

$$\frac{·1416}{·2146} = \frac{1 - 4a}{a} = \frac{1}{a} - 4.$$

That is to say $\frac{·1416}{·2146}$ is 4 less than $\frac{1}{·2146}$. We could substitute for ·2146 any number that would give a 4 in the quotient, and still obtain the same result—that is, any number greater than ·2 and less than ·25. So there is no particular virtue in the number ·2146.

In the second part we are finding:

$$\frac{·1416}{·2146} \div \frac{1}{·2146} = ·1416,$$

and if we ignore the decimal point we get the 1416 of the circle squarers—of course!

Another of these relationships is:

$$4\,\frac{·1416}{·2146} = 1·0000 \text{ or unity.}$$

Writing $a$ for ·2146, we have:

$$4\,\frac{1-4a}{a} = \frac{4a+1-4a}{a} = \frac{1}{a},$$

and if we then ignore the $a$ we get one, as we should from other numbers besides 2146.

One of the oddest things about these calculations is that 1416 is treated as if it were a separate entity. The value of $\pi$ is 3·1416 to a good approximation; but we cannot take a bit of this number and treat it separately in the way the circle squarers do. The separation into 3 and ·1416 is quite arbitrary; it has no more validity than writing the number as 3·14 and ·0016 or as 3·1 and ·0416, or indeed as 2·9 and ·2416. We even find the strange calculation:

$$\sqrt{3} = 1·732 + \sqrt{·14} = ·374 + \sqrt{·0016} = ·04 = 2·146.$$

Apparently we omit the decimal point and get 2146. All that can be said about that is that it is a minor oddity with no particular arithmetical significance.

Here is another strange calculation:

$$\tfrac{1}{7} = ·142857\ 142857\ 142857\ 142857$$

$$\begin{aligned}
\text{(three)} \quad &·1416 \\
&125664 \\
&\quad 502656 \\
&\qquad 2010624 \\
&\qquad\quad 8042496 \\
&\qquad\qquad 32169984 \text{ etc.}
\end{aligned}$$

We get each line from the one above it by multiplying by 4 and

going out four places. If we add up the series we get ·142857...,
'which shows clearly that the pi-ratio is absolutely definite and
final'.

What we have actually done is to multiply 3·1416 by the series:

$$1 + \frac{4}{10,000} + \frac{4^2}{10,000^2} + \text{ etc.}$$

and then to suppress the 3.

The sum of the series is:

$$\frac{1}{1 - \frac{4}{10,000}} = \frac{10000}{9996}.$$

And $\qquad 3·1416 \times \dfrac{10,000}{9996} = \dfrac{31416}{9996} = 3\,\dfrac{1428}{9996} = 3\tfrac{1}{7}.$

All we have done is to multiply 3·1416 by a factor that will change
it into $3\tfrac{1}{7}$.

If we are working backwards to find the series, we have:

$$3·1416 \times \frac{1}{1-r} = 3\tfrac{1}{7}.$$

$$r = 1 - \frac{3·1416}{3\tfrac{1}{7}}$$

$$= 1 - \frac{3·1416 \times 7}{22}$$

$$= 1 - ·9996, \text{ or } ·0004.$$

Hence we multiply by 4 and go out 4 places.

As distinct from these spurious coincidences there are some real
ones; it would be surprising if there were not. The equatorial
diameter of the earth is usually taken as $7926\tfrac{2}{3}$ miles. It turns out
that this is equal to:

$$\sqrt{2 \times 31416000} \text{ miles.}$$

So that we have twice the $\pi$ series under the root sign. Also if we
take the chord from pole to equator its length is $\sqrt{31416000}$ miles;
but that is the same coincidence, since we merely cancel $\sqrt{2}$.
Certainly an odd coincidence; but just one of those things that
happen. We have a choice of processes, and a choice of numbers;
and if one does not fit we try another.

Here is another coincidence:

$$\frac{\sqrt{3\cdot1416}}{2} = \cdot8862279$$

and

$$\frac{3\cdot1416}{28} = \cdot1122$$

$$\text{add } \frac{1}{2000} \times 3\cdot1416 = \quad 15708$$

$$\text{add } \frac{4}{10,000,000} \times 3\cdot1416 = \quad\quad 125664$$

$$\overline{\cdot11377205664}$$

Subtract from 1, and we get back to ·8862279.

Notice that the process is quite arbitrary. We add three fractions of 3·1416, and we choose the fractions that give the result we are looking for. In the reverse direction we want to find:

$$1 - \cdot8862279 = \cdot1137721.$$

We want the smallest number that will divide into 3·1416 and give a number not bigger than ·1137721. This number is 28.

$$3\cdot1416 \div 28 = \cdot1122,$$
$$\cdot1137721 - \cdot1122 = \cdot0015721.$$

We deal in the same way with ·0015721. The number is 2000.

$$3\cdot1416 \div 2000 = \cdot0015708,$$
$$\cdot0015721 - \cdot0015708 = \cdot0000013.$$

The 13 suggests multiplying by 4 instead of dividing; and we come to $\frac{4}{10,000,000}$ of 3·1416 = ·00000125664.

The most naive of the calculations is as follows: the speed of light is 186,300 miles per second. If we call it 186,283·2 then this is twice 93,141·6. Apart from the 9 we have the $\pi$ figures. How did they get there? By putting them there; the 283·2 added to 186,000 is responsible for all the $\pi$ figures except 31.

A great many people regard certain numbers as lucky or unlucky, usually unlucky. The favourite, or ill-favourite, is probably thirteen. We have all had experience of the ill luck attached to that number; it would be a strange thing if we had not. So many unfortunate things happen in the course even of a day; and we are surrounded by so many small numbers if we choose to look for them. The number of people in a room may vary from time to time, and we may choose any particular moment for a count. We can count the chairs in a room, the panes in the windows, the apples on a plate, and so on almost indefinitely. It would be strange if some of the counts did not occasionally reach thirteen; and it would be very strange if some of the numerous misfortunes did not coincide with thirteen. Of course we have to ignore any misfortunes not connected with thirteen, and equally we have to ignore any good fortune connected with thirteen. This kind of argument is called 'argument by enumerating instances'.

And of course it is flattering to our vanity to have something other than ourselves on which to blame our misfortunes. There are thirteen fruit trees in the garden; that is an excuse for failure, and not our own bad husbandry. Thirteen onions come up, instead of the hundred and fifty we expected; and even the thirteen are poor specimens. We blame the number, and forget our own bad composting, and lack of care in preparing the onion bed.

Give a number a bad name, and you can hang any misfortune on it.

As an experiment take the number eleven. We can begin by calling it 'the cheat's dozen'; we can remember the 'dark eleventh hour'. We watch day by day for misfortunes that can be attached to eleven. Any misfortune that happens on the eleventh of the month, or even on the twenty-second, can be ascribed to eleven. If there are eleven people at dinner, any misfortune that happens next day, or in the next week or month, or even in the next year, can be confidently ascribed to the malign influence of eleven. Very soon we shall have enumerated sufficient instances from our own experience to establish eleven firmly as an unlucky number.

Quite possibly one-seventh of all the ships that were wrecked sailed on Friday.

There is a lady of my acquaintance who is beset by a psychologist. He asks her to name a number at random; she says four. Why did she say four? He questions, and questions, till he discovers that she passed four traffic lights. Ha, he says. Now, was that the only number she had met that morning? She might have thought of her two boys, or five people in the house, or the ten fingers she carries with her wherever she goes. Or, oddly enough, she might just have thought of a number.

One of the troubles about numbers, especially when they come into stories, is that if you use a number at all it must be a particular number. It must be four, or five, or six, or seven, or some other number. Then, if the story is important enough, the mystery merchants get to work on it. Four may be the four cardinal points, and by implication all directions, the whole world, or even the whole universe. Or four may be the four sides of a square, their equality representing justice and uprightness. Or four may be four feet, and four may stand for the beasts as distinct from man. You choose whichever interpretation seems to your questing fancy to be most appropriate for the purpose you have in hand. If the number happens to be five, you probably elect for the five senses, even though there are more than five.

There is a perfectly straightforward story in the fourth chapter of *St John* in which Jesus meets a woman of Samaria by a well, and he talks to her. In the course of the talk Jesus said to her: 'thou hast had five husbands; and he whom thou now hast is not thy husband.' The gloss put on the number five, and extracted from Philo, is that the five husbands are really five seducers, and represent the five senses. For myself, I hardly see what Philo would make of the last statement 'he whom thou now hast is not thy husband'. But people who make mysteries out of numbers seldom trouble their heads with such trifles.

Strange perversity that when numbers are obviously meant to be read poetically they should be interpreted literally. The *Revelation* is full of numbers, and especially of the poet's number, which is seven. There are seven churches, seven candlesticks, seven stars which are the angels of the churches, a book with seven seals, seven horns and seven eyes, seven angels with seven trumpets, seven plagues and seven golden vials of wrath, a red dragon with seven heads and a beast with seven heads. Seven, seven, seven; the poet's number, crying out against literal interpretation. The twelve tribes of Israel introduce the number twelve; twelve thousand are sealed from each tribe, making an hundred and forty and four thousand from all the tribes: twelve times twelve in thousands. The Holy City has twelve gates, and the wall of it has twelve foundations; the height of the wall is twelve times twelve cubits; the length and breadth and height of the City are twelve thousand furlongs. The number of the elders is twice twelve. Twelve, twelve, twelve; another number crying out in vain against literal interpretation. There are four beasts each with six wings (twice twelve wings), four horsemen, four angels standing at the four corners of the earth holding the four winds, four angels bound in the great river Euphrates. Those are the fours.

The times are sometimes definite: five months, a thousand two hundred and threescore days, a thousand years. And sometimes vague: an hour, and a day, and a month, and a year; a time, and times, and half a time. The fractions are nearly all thirds.

There is a sudden jump when we come to big numbers: ten thousand times ten thousand; and bigger still the two hundred thousand thousand horses by whom were a third of men killed. And lest any one should think that the inhabitants of the Holy City were few the poet cries out 'I beheld, and, lo, a great multitude, which no man could number, of all nations, and kindreds, and peoples, and tongues.'

I have sorted out the numbers to draw attention to their quality. The analysis leaves no excuse for any interpretation except the obvious one. If anyone is still inclined to attach magical properties to seven and twelve, he might remember that these numbers have come down in a long line of credulity from ancient Babylon.

The famous 'number of the beast' 666 comes from the Jewish method of cabalistic interpretation by assigning certain numbers to the letters of words. 666 is the number of Emperor Nero when written in Hebrew characters, according to the art of gematria as it is called.

In *War and Peace* Tolstoy gives the numerical equivalents of the letters:

| a | b | c | d | e | f | g | h | i | k | l | m | n |
|---|---|---|---|---|---|---|---|---|---|---|---|---|
| 1 | 2 | 3 | 4 | 5 | 6 | 7 | 8 | 9 | 10 | 20 | 30 | 40 |

| o | p | q | r | s | t | u | v | w | x | y | z |
|---|---|---|---|---|---|---|---|---|---|---|---|
| 50 | 60 | 70 | 80 | 90 | 100 | 110 | 120 | 130 | 140 | 150 | 160 |

L'Empereur Napoléon has the number:

$$20+5+30+60+5+80+5+110+80+40+1$$
$$+60+50+20+5+50+40=661;$$

so if we add 5 for the *e* dropped from the *le* before *Empereur*, we arrive at 666 for Napoleon's number; and this left no doubt of the identification of Napoleon with Antichrist.

In Tolstoy's satire of the art of gematria even further manipulation is necessary. Besouhoff respells his name Besuhof, puts le russe in front of it, and finally elides the *e* before russe. So that L'russe Besuhof gives the number 666; and there he was.

It is commonly said that statistics can be made to prove anything; and there is a small element of truth in the saying. The fact is that the unscrupulous manipulation of numbers can be made to give any result that the manipulator aims at. Hitler's number, for example, is 222, which is a third of 666; and this proves conclusively that he is but a third of a man or beast, or alternatively that he has two doubles. If we take his original name of Schicklgruber, his number is 427. If we divide by the poet's number, we get 61. So evidently Schicklgruber is to end in his 61st year, with no poet to weep him, honour him, or sing him. If we take the other number:

$$222=37\times6,$$

so that evidently his range of power is for six years after (19)37, and this brings his end in 1943. Some prophets would have suppressed this prediction, and embarked on the easy task of making another. I leave it as a warning.

Bismarck's number is 225, and if we omit the unnecessary *c*, it is 222, so that evidently Hitler is a reincarnation of Bismarck, probably with some quality omitted. 3 stands for reason and opinion, so apparently these qualities are omitted.

Alas, divination is as old as credulity; and the fraudulent manipulation of numbers can look impressive—until the method of manipulation is analysed.

# CHAPTER 16

## *Was Man made by the Earth?*

OR ALTERNATIVELY, was the earth made for man?

There are at least two possible answers. Many biologists are inclined toward the easy answer: that certain conditions happened, fortuitously, to be present on the earth, and that life adapted itself, more or less successfully, to these conditions. It is not surprising that biologists should be inclined to take this view; biologists are constantly in contact with examples of adaptation, some of them very strange.

Most of the stress has been on adaptation to environment. What is less often considered is the nature of the environment. There are very severe limits to adaptation; the fact that we do not often come up against these limits merely shows how nicely conditioned to life the earth is.

We can start with the fact that the materials of the universe seem to be the same everywhere. By reproducing temperature and pressure conditions we confidently expect the same reactions and linkages in one place that would occur in any other place.

To imagine a sort of solid crystalline life, or life in an atmosphere of cyanogen, as some have pretended to imagine, is mere folly. Such a pretence ignores all the relevant elements of the case: the nature of life, the nature of crystals, and the nature of cyanogen. It ignores not merely the fact that cyanogen is a poisonous gas, but the reason why it is poisonous.

The next fact relevant to environment is the extraordinary nature of the solar system. It is now commonly ascribed to a glancing blow from another star which flung out or drew out matter from the sun, to condense at intervals into planets.

The odds against such an occurrence are considerable, though not nearly so high as was once thought. It is estimated that there is one star in every ten parsecs of space, so that the average distance between stars is $\sqrt[3]{10}$ parsecs. For the present purpose we can take this as two parsecs, or about 40 billion miles. The sun is estimated to be moving at a rate rather less than 5000 million miles per year, relative to surrounding stars, so that it would traverse the distance between star and star in:

$$\frac{40 \text{ billion}}{5000 \text{ million}} = 8000 \text{ years.}$$

The sun has the whole of a 2 parsec square to aim at, as it were, with the next star occupying some point on the square. The area of the square is:

$$(40 \text{ billion})^2 = 1600 \times 10^{24} \text{ square miles.}$$

It has been calculated that the glancing star must have approached the sun within a distance of three radii in order to produce the necessary tidal action. In the present state of the sun this would mean an approach to within $1\frac{1}{2}$ million miles. The area of a circle with this radius is:

$$\pi \times (1\frac{1}{2} \text{ million})^2 = \text{about 7 billion square miles.}$$

This is:

$$\frac{7 \times 10^{12}}{1600 \times 10^{24}} \text{ of the area of a square parsec}$$

$$= \text{about } \frac{1}{200 \times 10^{12}} \text{ of that area.}$$

So there is one chance in 200 billion every 8000 years of the sun coming close enough to another star to have the necessary effect in producing a system similar to the solar system. In the 2000 million years or so of the sun's existence in its present state, the chance would recur:

$$\frac{2000 \times 10^6}{8000} = 250{,}000 \text{ times.}$$

The odds against the occurrence are thus reduced to:

$$\frac{200 \times 10^{12}}{250{,}000} \text{ to one}$$
$$= 800 \times 10^6 \text{ to one.}$$

That is, the odds against such an occurrence during the lifetime of the sun in its present state are 800 millions to one.

Before the sun condensed it had a comparatively brief existence in an expanded state, a period of a mere 20 million years. Other things being equal there would have been 100 times less chance of a collision in this short period than in the 2000 million years since the sun condensed. But the radius in the expanded state may have been as much as 100 million miles. The target area would be increased by a factor of:

$$\left(\frac{100 \text{ million}}{\frac{1}{2} \text{ million}}\right)^2 = 200^2 = 40,000.$$

Taking the two results together we find that there was $\frac{40,000}{100} = 400$ times more chance of a collision in the early expanded period than in the whole of the later condensed period. Most of the latter period must be ruled out, to allow time for the planets to condense and cool, and for the earth to go through the long processes of geological and biological change. It becomes highly probable that the highly improbable collision occurred in the early period when the expanded sun occupied much more space than now.

Though planetary systems are probably unusual, there seems to be little doubt that there must be many of them amongst the hundred thousand million stars of the galactic system. However, the mere existence of some kind of planetary system does not necessarily mean the possibility of life. There are many other things that have to be taken into account before we can consider life a possibility.

We have to take into consideration the size of the earth; and that is an extraordinary circumstance in the environment of man. The earth appears to be the right size for animals, and in particular for upright man. If the earth had twice the diameter, and if it were made of the same materials, it would have eight times the volume and mass. Weights on the surface of such a planet would be twice as great as on the earth $\left(\frac{8}{2^2} = 2\right)$. We have all felt the effects of very short-lived changes of weight; but it is difficult to imagine the effects of a large permanent increase. If weights were doubled, the heart would have to pump blood that was twice as heavy, and we should need a different kind of heart. The muscles of a 10-stone man would have to support and manipulate permanently an additional weight of 10 stones; he would certainly need extra muscles and other tissues to bear part of the burden. There would have to be very severe adaptations to meet the stress of such conditions; and we can rule out the possibility of any creature with the grace and upright carriage of man.

So far as size is concerned Venus is the same kind of planet as the earth, and in a different position it might be a fit home for men. Jupiter apparently has not condensed sufficiently to be a planet with a solid surface capable of sustaining life. It has 314 times the mass of the earth, and only a quarter of the density. If it were to shrink to such an extent as to have the density of the earth, its diameter would be:

$$\sqrt{314} = 6{\cdot}8 \text{ times the diameter of the earth.}$$

Materials on the surface of such a planet would weigh 6·8 times as much as on the earth. Adaptability could hardly stretch so far as to enable life to function in such conditions. We have to imagine the toughest of tough skins (made of what materials?), diminutive size, heavy water that finds it difficult to evaporate, the small life-giving forces crushed out of recognition by mere weight, and a million other impossible conditions. He would indeed be an ardent adaptationist who would attempt the profitless enterprise of imagining life on a large planet.

On a smaller planet than the earth conditions as regards weight would be easier, and plants and animals might grow to a greater size. But other difficulties would arise. Water would evaporate more easily, so that it would be more difficult to have liquid water. And a small planet would be unable to retain much atmosphere, even if any at all. Mercury seems to have lost its atmosphere, including water; so also does the moon. Mars is a small planet with a thin atmosphere, and probably little water. Whereas the earth and Venus are of a size to retain a dense atmosphere and abundant water.

When we think of the earth as a habitation for man, and not merely as a place where some sort of rudimentary life could exist, we begin to look for some degree of comfort in the environment—the sort of comfort that would make human life a reasonable possibility. The rate of rotation of the earth seems to be one such comfortable condition. A really uncomfortable extreme in one direction would be the kind of rotation the moon has, that is rotation in the same period as revolution. It seems probable that Mercury rotates in the same time that it revolves round the sun. If so, one side is perpetually exposed to the heat and glare of the sun, perpetually accumulating heat with no cool nights to ease the temperature. The other side has perpetual night, with never a glimpse of the sun, and is for ever frozen.

An uncomfortable extreme in the other direction would be a speed of rotation that would cause things to fly off from the earth. The acceleration outward due to rotation is given by $\frac{v^2}{r}$. $v$ is the velocity of a point on the earth's surface; $r$ is the radius of the circle in which it is rotating. We want to find $v$ at the point where $\frac{v^2}{r} = g$, so that the outward acceleration is equal to the inward acceleration. At this point objects have no weight, no downward pull toward the earth, and they begin to fly off. For convenience we will consider a place on the equator; $r$ is the radius of the earth = 3963 miles = 3963 × 5280 feet.

We want: $\frac{v^2}{r} = g$ or $v = \sqrt{rg}$.

$$v = \sqrt{3963 \times 5280 \times 32} \text{ feet per second}$$
$$= \tfrac{15}{22} \sqrt{3963 \times 5280 \times 32} \text{ miles per hour}$$
$$= \text{about } 17{,}600 \text{ miles per hour,}$$

and that is more than 17 times the speed at which a point on the equator is rotating (about 1000 miles per hour). That is, if the earth were to rotate in 1½ hours things would begin to fly off, including the atmosphere. Long before we come to such a disastrous speed of rotation we reach a speed that would be very uncomfortable. Short nights and short days would leave insufficient lengths of time for work and play and sleep. Quite possibly life might adapt itself to much shorter days and nights, but not comfortable human life.

A considerable slowing down of rotation would bring in another series of uncomfortable results. Four or five extra hours of daylight would mean that much extra time for receiving heat from the sun, and a great increase in the afternoon temperature. In the night there would be four or five extra hours during which heat would be radiated into space, and a considerable fall in the early morning temperature. We should have, in an exaggerated form, the uncomfortable conditions that actually exist in desert climates: extremely hot days followed by extremely cold nights.

We are sometimes surprised at the conditions in which plants and animals manage to exist; but the range of temperature within which life is possible is actually very narrow. The whole possibility of temperature runs from the $-273°$ C. of absolute zero, up to the millions of degrees that has been calculated for the interiors of stars. Out of that vast range of temperature life is restricted to the narrow interval between $0°$ C. and perhaps $40°$ C. That is to say, the temperature must not be permanently so low that all water is perpetually frozen; and it must not be permanently so high that it is anywhere near the boiling point of water. Within that narrow range of temperature life can adapt itself in various forms. It is no real objection to this argument that life can exist temporarily in frozen conditions. If the conditions in any region are permanently frozen, life can only exist there at the expense of unfrozen parts of the world.

That is why the earth had to be just the distance it is from the sun. Venus, our nearest neighbour amongst the planets, is ·723 astronomical units from the sun, that is ·723 times the distance of the earth. So it receives:

$\dfrac{1}{·723^2} = 1·9$ times as much light and heat from the sun  as the earth does.

That is so vast an amount that it is doubtful whether water could exist there as a liquid, at any rate during the day; there might be heavy deluges at night. The atmosphere of Venus appears to be laden with clouds—that is one reason for the brilliant appearance of the planet; the presence of the clouds indicates a large amount of water vapour in the atmosphere.

Mars is half as far away again as the earth, so it receives:

$$\frac{2^2}{3^2} = \frac{4}{9},$$

or less than half the amount of heat and light the earth receives. If there is any life on Mars, it must be very rudimentary. A planet with a temperature which seldom rises above the freezing-point of water is hardly a fit home for men.

The further out from the sun a planet is situated the longer is the time it takes to go round the sun. And therefore, also, the longer are its seasons if its axis happens to be inclined. The inclination of Mars' axis is a little more than that of the earth's axis. It takes nearly twice as long to go round the sun as the earth does, so that its seasons are nearly twice as long. A twelve months winter must be a pretty desperate affair. People who write, light-heartedly, about something like human life on Mars should really pay more attention to the nature of Martian seasons and especially to the Martian winter. They should set themselves to think of a planet receiving a mere half of our light and heat, and doomed to a winter extending over eleven or twelve months.

H. G. Wells made a brilliant attempt to imagine life on the moon, with a fortnight of blinding day, followed by a fortnight of devastating night. As the high priest of adaptability he pushed that doctrine to the limit of *reductio ad absurdum*; no one has lampooned adaptability so brilliantly as he. He pushes the idea of adaptability to the extreme limit of imagining plants growing so quickly, to meet the stress of a mere fortnight of spring, summer and autumn, that the rate of growth can be seen. One has only to compare the nightmare Wellsian idea of adaptability with the real thing, to perceive how restricted real adaptability actually is.

Thus we have the general condition of the solar system as at least unusual. We have in the earth a planet of the right size to make human life possible. We have this planet situated at the right distance from the sun, so that it receives enough light and heat, but not too much; and so that its seasons are long enough, but not too long. In addition we have the planet rotating at the right speed, so that day and night are long enough, but not too long. That is an extraordinary concatenation of favourable circumstances. One can ascribe it to chance, or, perhaps mor reasonably, to design.

Nevertheless, we might have had all these favourable circumstances concentrated on the earth, and still the earth might have been completely inert if it had not been for another set of circumstances even more extraordinary. These circumstances involve the properties of some of the elements and their compounds.

Water is one of the prime necessities of life, and it is a very extraordinary liquid. It would hardly be stretching things too far to say that it is a liquid designed for the earth; at least it is exactly suited to earthly conditions. It exists as a liquid within a very restricted range of temperature, 0° C. to 100° C. in normal earthly conditions. It readily evaporates and readily freezes, so that on the earth we have the advantage of all three forms: solid, liquid and gas. Arithmetically, water is an earthly substance, exactly suited to our conditions. Add to this that it is a thing of beauty in itself, and that it adds beauty to landscapes that would otherwise be harsh and angular. It is probably true that nearly all natural beauty is associated with water.

One of the most extraordinary things about water is its contraction with increase of temperature from 0° C. to about 4° C. We call such behaviour anomalous, a variation from an all but universal rule. The contraction, and consequent increase in density, is very small; the density increases from ·99987 at 0° C. to 1 at 4° C. The increase in density is 13 parts in 100,000 or about 1 part in 7700. Even this small increase in density has very important results. When water is cooling the heavier water sinks to the bottom. The heaviest water of all is water at 4° C. above freezing point; so that below 4° C. colder water no longer sinks. It is for this reason that ponds and lakes freeze from the top downward, so that there is nearly always liquid water below the ice. That is why fishes can survive a severe winter without undue hardship.

Another extraordinary thing about this extraordinary liquid, water, is its high specific heat; its specific heat is higher than that of almost any other substance. It takes a lot of heat to raise a mass of water through a given range of temperature; and, once the water has been raised in temperature, it has the same large amount of heat to give out. A pound of sand, for example, would take a mere fifth of the amount of heat to raise it from 0° C. to 100° C. that would be required to raise the same amount of water through the same range of temperature; and of course it would give out in cooling only a fifth of the amount of heat that water would give out. That is why the oceans can retain a vast amount of heat, and so can mitigate extremes of climate by slowly giving up the heat.

Another extraordinary thing about water is that it dissolves an enormous number of substances, and in particular that it readily dissolves air. That is why fishes can live in it.

The earth is fortunate in having so vast a store of so convenient a liquid as water; it is further fortunate in having this liquid stored in the deep basins of the oceans. We have seen that the water of the oceans forms more than a thousandth part of the whole mass of the earth. Who but an omniscient creator would ever have dreamed that so vast a store of water was necessary to life—that the comparatively diminutive continental masses must be balanced by masses of water much greater than themselves? Even as it is, we have deserts in the arid hearts of continents, and we have frozen lands in places where the watery amelioration is not felt. What an amateur creator would make of the world it is difficult to think; though we can be sure that it would be an unholy mess.

The atmosphere of Mr Wells' moon is an odd sort of thing. It seems to freeze in the winter night. In spite of this there is a flow of air from the frozen side to the summer side. The terrestrial atmosphere is a very different thing from this imaginary moon atmosphere. It obeys natural laws; it does not freeze, for example, except when the necessary conditions of temperature and pressure are produced artificially. The terrestrial atmosphere is a remarkable mixture of nitrogen and oxygen, four parts of nitrogen to one part of oxygen; the remarkable thing is that so much nitrogen should be necessary. An atmosphere of pure, or nearly pure, oxygen, would turn every flame into a conflagration; it is the dilution with nitrogen that enables things to burn without burning too fiercely. The presence of free oxygen in the terrestrial atmosphere, suitably diluted with a much larger proportion of harmless nitrogen, is one of the most astonishing things in this astonishing planet.

Perhaps more astonishing still is the presence of a small amount of carbon dioxide in the atmosphere. The exact amount is important, because it is one of the conditions of life. The amount is no more than 4 parts in 10,000. That is not nearly enough to interfere with either breathing or burning; nevertheless, it is vitally important. The atmosphere actually contains a large amount of carbon. I want to draw attention to the amount, because carbon is the basis of the compounds in which life exists.

The weight of the atmosphere is 14·7 pounds per square inch of the earth's surface. This area is 197 million square miles:

$$= 197 \times 10^6 \times 5280^2 \times 12^2 \text{ square inches};$$

so the total weight of the atmosphere is:

$$14\cdot7 \times 197 \times 10^6 \times 5280^2 \times 12^2 \text{ pounds}$$

$$= \frac{14\cdot7 \times 197 \times 10^6 \times 5280^2 \times 12^2}{2240} \text{ tons}$$

$$= 5\cdot2 \times 10^{15} \text{ tons,}$$

or about 5000 billion tons.

Of this total weight 4 parts in 10,000 is carbon dioxide, or $20\cdot8 \times 10^{11}$ tons

$$= 2 \text{ billion tons.}$$

Carbon dioxide contains 12 parts of carbon in 44 parts; so the weight of carbon in the atmosphere is:

$$\tfrac{12}{44} \times 20\cdot8 \times 10^{11} = 5\cdot7 \times 10^{11} \text{ tons,}$$

or over half a billion tons.

That is the weight of atmospheric carbon that is available for the growth of plants, and therefore for the sustenance of both plant and animal life. It is a large amount, but not excessive as a constituent of the atmosphere.

In addition to atmospheric carbon there is a thin layer of carbon tied up in trees and other plants that cover the earth, and in animals. One of the great equatorial forests may carry 100 tons or more of carbon to the acre. The Amazonian forest covers an area of a million square miles. The amount of carbon on that million square miles must be in the region of:

$$10^6 \times 10^2 \times 640 \text{ tons}$$

$$= 64,000 \text{ million tons.}$$

This is a considerable fraction of the amount of atmospheric carbon. The fraction is:

$$\frac{64,000 \times 10^6}{570,000 \times 10^6} = \text{about } \tfrac{1}{9} \text{ of the amount.}$$

Anything like tropical vegetation over the whole land surface of the globe would be impossible. The half billion tons of carbon in the atmosphere would provide something like $2\frac{1}{2}$ billion tons of trees, or perhaps a billion trees. Allowing 60 trees to the acre, these trees would cover:

$$\frac{10^{12}}{60 \times 640} \text{ square miles}$$

$$= \text{about } 26,000,000 \text{ square miles.}$$

So that in the highly improbable event of all the carbon being withdrawn from the atmosphere we could have another 26,000,000 square miles of forest.

It should be clear that so far as the supply of available carbon is concerned we are running on a rather narrow margin. The supply of carbon appears to be ample, but not excessive.

We may note in passing that the burning of a thousand million tons of coal is no inconsiderable thing. It adds one five-hundredth part to the supply of atmospheric carbon. That is the only thing I have ever heard of or thought of in favour of the rapid burning up of coal supplies.

I often wonder if organic chemists are as excited as they might be about the exciting element they are dealing with. The properties of the carbon atom are truly extraordinary. If these properties are a fortuitous circumstance, they are just about the most extraordinary accident it is possible to imagine. We can list the linkages of other elements, but the linkages of the carbon atom are a whole subject in itself. Carbon compounds are literally innumerable.

It is the four valencies of the tetrahedral carbon atom that enable it to link up into innumerable compounds. The carbon atom has four links, and it has the unusual property of using one or more of the links to join up with other carbon atoms. This property is exhibited most remarkably in the diamond, in which each carbon atom is chemically linked to four other carbon atoms situated tetrahedrally around it. Each diamond is a single molecule, a 'giant molecule', in which each atom is linked to four other atoms of carbon. It is these chemical links that put the hardness of the diamond in a class by itself.

Usually the carbon atoms link up incompletely with other carbon atoms in chains or rings. Two or three of the links are thus left free to join up with other atoms or groups of atoms. It is in this way that the multitudinous carbon compounds are formed; and it is in association with these compounds that life can exist. Without the tenuous carbon covering of the earth, and the half billion tons of carbon in the atmosphere, the conditions of life would not exist, and the earth would be dead.

Water, air, and carbon, plus a few other elements, are not the only things necessary to life. There has to be the capability of using them; and that involves life itself and a number of small forces that are the mechanism of life. There is the extraordinary process by which plants manufacture starch from carbon dioxide and water. There are the forces of surface tension and capillarity, osmosis, adsorption, and similar forces. These forces are small, but adequate in amount to enable sap to rise to the top of the highest tree; they function on small but adequate margins.

Adaptability can be made to account for a lot of change and development inside the framework of terrestrial environment. The things it does not account for are: the origin and nature of life itself; the existence of a group of small forces that form a mechanism for life and so enable it to function; the remarkable properties of water, atmospheric oxygen, and the carbon atom; the extraordinary distribution on the earth's surface of water, air, and carbon; the concentration on the earth of circumstances favourable to life, and especially to the life of man; the narrow margins within which the forces of life function.

My own answer to the double question I have propounded is: only to a small extent has man been moulded by terrestrial environment. Arithmetically, the earth is the Lord's.

Printed in the United States
By Bookmasters

Printed in the United States
By Bookmasters